Edward Olney

A Primary Arithmetic and Teacher's Manual

With Class and Seat Exercises graded with Reference to the various Stages of the

Pupil's Advancement in Reading

Edward Olney

A Primary Arithmetic and Teacher's Manual
With Class and Seat Exercises graded with Reference to the various Stages of the Pupil's Advancement in Reading

ISBN/EAN: 9783337179724

Printed in Europe, USA, Canada, Australia, Japan

Cover: Foto ©ninafisch / pixelio.de

More available books at **www.hansebooks.com**

A

PRIMARY

ARITHMETIC,

AND

TEACHER'S MANUAL,

WITH

CLASS AND SEAT EXERCISES GRADED WITH REFERENCI
TO THE VARIOUS STAGES OF THE PUPIL'S
ADVANCEMENT IN READING.

By EDWARD OLNEY,

PROFESSOR OF MATHEMATICS IN THE UNIVERSITY OF MICHIGAN, AND AUTHOI
OF A SERIES OF MATHEMATICAL TEXT-BOOKS.

NEW YORK:
SHELDON & COMPANY.
1876.

OLNEY'S SERIES OF MATHEMATICS.

OLNEY'S PRIMARY ARITHMETIC.

OLNEY'S ELEMENTS OF ARITHMETIC.

OLNEY'S SCIENCE OF ARITHMETIC. (*In preparation.*)

TEACHER'S HAND-BOOK OF ADDITIONAL EXAMPLES AND EXERCISES. (*In press.*)

INTRODUCTION TO ALGEBRA $1 00

COMPLETE SCHOOL ALGEBRA 1 50

TEST EXAMPLES IN ALGEBRA 75

OLNEY'S HIGHER MATHEMATICS.

UNIVERSITY ALGEBRA 2 00

ELEMENTS OF GEOMETRY 1 50

ELEMENTS OF TRIGONOMETRY 1 50

GEOMETRY AND TRIGONOMETRY, UNIVERSITY EDITION. 3 00

GENERAL GEOMETRY AND CALCULUS 2 50

PREFACE.

IT is thought that the spirit of this book, and the manner of using it, will be so evident, as the teacher reads it over, that few prefatory words are needed. The following are some of the leading principles by which the author has been guided. How they have been wrought out can be seen only by examining the book itself.

1. A text-book for schools should be arranged with reference to sound principles of teaching, and to convenience of use in the school-room, quite as much as to the principles of the science which it develops.

2. *One thing at a time* is the fundamental maxim of primary teaching. Each exercise must have a single, clearly defined purpose.

3. *Unity of purpose* and almost infinite *diversity of means* characterize the most successful teaching of the young.

4. The young child must be furnished something to *do*. His hands, his eyes, and, as much as may be, his tongue and his whole body, must be busied with the work in hand.

5. In a well-conducted primary school, as careful attention will be given to secure profitable employment for the pupils in the seats, as to the conduct of the class exercises.

6. The cases are exceptional, and very rare, in which much labor or time need be bestowed in order to awaken in the mind of the child the conception of number. The recognition of number is one of the most simple, earliest developed—in fact,

most nearly innate—of all our mental acts. The child who has learned to count 10 by means of objects, has as well-defined, practical notions of number, as he needs, and as he advances with the simple processes of combination, his conceptions will enlarge as occasion requires.

7. There are two distinct mental processes required in obtaining a mastery of the elementary combinations of numbers:

1. The method by which we determine what the result of the combination is, and

2. The fixing of that result in the memory.

Thus, it is one thing for the pupil to learn how he may find out how many 6 times 7 makes, and quite another thing to fix this product in his mind. The former is a *process*, which the child who can count will readily learn, and which he will always apply with pleasure. The latter is a pure act of memory, and the pupil needs all the help an ingenious teacher can devise, to save it from becoming intolerable drudgery. In a single lesson, the child who can count one hundred, will learn to make the Multiplication Table as far as 10 times 10. But to remember these 100 products, so that they can be instantly named, is no less a task than to memorize the answers to any other 100 problems. The same may be said of the Addition, Subtraction, and Division Tables; for they are none of them well learned until the results can be recalled without any mental process except the instantaneous act of the memory.

8. *To perceive* and *to remember* are the chief mental exercises of the grade of pupils for which this book is prepared. Such pupils cannot be expected to give formal statements either of definitions, processes, or reasons; and much less can they obtain conceptions and learn processes from abstract statements. Hence, formal definitions, rules, and processes of reasoning are out of place in such a book.

9. Usually the child who cannot count cannot read; and the processes of learning to read and learning the elementary com-

binations of numbers are going on at the same time. Accordingly, in this book the first 27 pages are addressed to the Teacher; the next 90 are addressed to "pupils reading simple words;" and the remainder of the book assumes that the pupil has learned to read tolerably well.

10. According to the decimal notation, the fundamental combinations embrace only numbers below and including 10. To such combinations this book is, therefore, confined.

11. One is more interested in what he has made himself, than in that which is furnished by another. Hence the pupil is taught *how to make* the Addition, Subtraction, Multiplication, and Division Tables for himself, and, having made them, *to study his own work*. None of these tables are given, except in form, in this book.

12. From objects in sight and in hand to objects out of sight —from the concrete to the abstract, from the known to the unknown, by short and easy steps—an arrangement which will make each advance include a practical review, etc., are principles so well established that no intelligent teacher will countenance the violation of them.

The teacher who is familiar with the methods of the Kindergarten will recognize the spirit of those methods on every page of this book. Indeed, it has been a leading purpose to embody this spirit in forms which are practicable for use in our ordinary Primary Schools.

EDWARD OLNEY.

UNIVERSITY OF MICHIGAN, *December, 1874.*

NOTE.—Since the plates of this book were first cast, the whole book, in complete form, has been thoroughly examined by a number of practical teachers in different parts of the country, and carefully revised. The exceedingly liberal spirit of the publishers has allowed the author to make such revision to any extent he desired. To Prof. N. A. CALKINS, of the New York City Normal School, the author is very greatly indebted for valuable suggestions in connection with this work of revision.   E. O.

INTRODUCTION.

ORGANIZATION AND EXERCISES OF A PRIMARY SCHOOL.

SO much attention has been given of late to primary teaching, and principles and methods have been so rapidly developed, that the author has thought that a synopsis of a few of these results would be acceptable to the teacher. The preface recites some of the established principles; it is the purpose of this introduction, and, in fact, of this book, to exhibit in outline an embodiment of such principles in method.

So great is the diversity among our Primary Schools, that it is practically impossible to present a schedule which is adapted to all. What is designed in this attempt is to indicate somewhat of the plan of organization and course of exercises found in our best Primary Schools in towns of 3,000 to 8,000 inhabitants, where the schools are graded into 4 or 5 departments. In the larger cities, where there are two or more primary grades, the oral exercises can be more frequent for each class, and still greater variety will be practicable. Nevertheless, the spirit and general features of the scheme may be much the same in all.

Such a school as is here described will consist of 50 or 60 pupils arranged in 3 classes, styled respectively the "A" class, "B" class, and "C" class, the first being the most advanced, and the last the least. The pupils of each class will be seated together, as seen in the cut on page 5, where the "A" class occupies the two forms at the left, the "B" class, which is at the counting table, occupies the two centre forms, and the "C" class the two right-hand forms.

The age of the "C" class will vary from 5 to 7, and pupils will be, on an average, about a year in each class. The "A" class will usually be found reading in what is called the "Third Reader," and will be able to learn easy lessons in descriptive subjects, as Natural History and Geography.

No exercise should occupy more than 15 minutes, and with the younger classes many of the class exercises need not exceed 5 or 8 minutes. Ten minutes are assigned to most of the separate exercises in the schedule. Time saved by the shorter exercises will give opportunity for inspection of work, singing, or any of the numerous, nameless things which need attention.

This plan supposes that the pupil will be kept constantly busy, recreation being as regularly provided for as work. "st." means "exercise in seat," "cl." means "class exercise," and "B—B" means work on the pupils' blackboard. The class exercises are printed in full-faced type.

Part of the writing exercises will be for the purpose of learning to write, and part for the purpose of learning to spell. The former will usually be from copy, and the latter from dictation by the teacher as she is about her other work.

The drawing exercises will comprise geometrical forms, tracing from copies, simple natural objects, and outline map drawing.

The oral exercises will be largely what is known as "Object Lessons." These will be on various subjects, such as color, form, common properties of bodies, direction, etc.

The "A," "B," and "C" classes as here designated correspond with the 1st, 2d, and 3d *Grades*, or years, respectively, of the system now coming into use in many of our Graded Schools. In the larger of these schools it is assumed that each grade will be divided into two *Divisions*.

PROGRAMME FOR A DAY IN A PRIMARY SCHOOL.

FORENOON.

TIME.	"A" CLASS.	"B" CLASS.	"C" CLASS.
9 to 9:15	Opening Exercises.		
9:15 to 9:25	Writing, st.	Reading, st.	Reading, cl.
9:25 to 9:35	Reading, st.	Reading, cl.	Arithmetic, st.
9:35 to 9:45	Reading, cl.	Drawing, st.	Printing, B—B.
9:45 to 9:50	Gymnastics, and Oral Concert Exercises.		
9:50 to 10	Drawing, B—B.	Arithmetic, st.	Arithmetic, cl.
10 to 10:10	Arithmetic, st.	Arithmetic, cl.	Drawing, st.
10:10 to 10:20	Arithmetic, st.	Writing, st.	Oral Teaching.
10:20 to 10:40	Recess.		
10:40 to 10:50	Arithmetic, cl.	Drawing, B—B.	Writing, st.
10:50 to 11	Geography, st.	Reading, st.	Reading, cl.
11 to 11:10	Geography, st.	Reading, cl.	Drawing, st.
11:10 to 11:20	Geography, cl.	Drawing, st.	Arithmetic, st.
11:20 to 11:30	Gymnastics, and Oral Concert Exercises.		
11:30 to 11:40	Writing, st.	Arithmetic, st.	Oral Teaching.
11:40 to 11:50	Arithmetic, st.	Arithmetic, cl.	Printing, st.
11:50 to 12	Arithmetic, cl.	Printing, st.	Drawing, B—B.

AFTERNOON.

2 to 2:10	Writing, st.	Reading, st.	Reading, cl.
2:10 to 2:20	Nt. History, st.	Reading, cl.	Arithmetic, st.
2:20 to 2:30	Nt. History, cl.	Writing, st.	Printing, B—B.
2:30 to 2:45	Gymnastics, Stories, and Moral Lessons.		
2:45 to 2:55	Arithmetic, st.	Arithmetic, st.	Arithmetic, cl.
2:55 to 3:05	Arithmetic, st.	Arithmetic, cl.	Writing, st.
3:05 to 3:20	Recess.		
3:20 to 3:30	Arithmetic, cl.	Drawing, st.	Drawing, st.
3:30 to 3:40	Drawing, st.	Spelling, st.	Oral Teaching.
3:40 to 3:50	Spelling, st.	Spelling, cl.	Printing, st.
3:50 to 4	Spelling, cl.	Writing, st.	Drawing, st.

SECTION I.

COUNTING, AND READING AND WRITING NUMBERS FROM ONE TO ONE HUNDRED.

THIS Section is addressed to the *Teacher*. It is presumed that pupils who cannot count, cannot read; and hence that the text of a book can be of no service to them. The pictures in this section will be useful to the pupils, as will appear in the progress of the lessons. Hence the pupils will need the book from the beginning.

The lessons of this section, however, will be wholly oral. We call this class of pupils the "C" class, the lowest grade in the Primary School. (See Introduction.)

Appliances.—1. A good **Blackboard**, crayons, rubber, and pointer, for the teacher's use. The blackboard should be about 3 feet by 9.

2. A **Table** about 3 feet by 6, so arranged that the top can be inclined towards the class, and low enough so that children of 5 or 6 years can see its surface, as they stand around it, and can get their hands on it conveniently.

3. **One Hundred Counters.** Tasty counters can be cut from bright colored, heavy card-board. They should be about ¾ of an inch square; or, if circles, about the same in diameter. Common wooden button molds will answer. Whatever is used should be neat and convenient, but so simple as not to attract undue attention. Small bundles of splints are much used.

4. A **Numeral Frame**—a necessity in a Primary School.

5. A **Long Blackboard** on the sidewall, so low that children of this age can write on it easily, and long enough for 18 or 20 pupils to stand before it at once, and write on it.

6. Each pupil needs a **Slate and Pencil.**

LESSON I.

Purpose.—*To teach to Count from One to Ten.*

Method.—**Class Exercises.** While the teacher stands behind the Counting Table, and the class is gathered around it, as represented in the picture, let the teacher have *ten* counters lying together on the table, and moving out *one* of them, ask, "How many is that?" When this is answered by all, move out another, and placing it with the first, ask, "How many have we now?" In this way see how far any of them can count. If all can count *ten,* readily, there is no need of spending more time on the exercises in this lesson. But *class* answers must not be depended upon for determining this—each pupil must be questioned separately, while all look on, and, when occasion serves, help.

If few, or none of them, can count ten, the work of teaching must be continued till all can count thus far, readily. Use the counters as above, having the pupils count *in concert,* at first, as the teacher moves out the counters. Then encourage *individuals* to try it. "Now, who can count four?" "Well, Jane, count out four of the counters." "Who else can do it?" "Who else?"

"Who can count out seven counters?" "Well, James, you may try it." "Who else?" etc., etc. Vary the exercise by having the pupils count marks, or dots, as you make them on the board. Also, if they know any letter, as o, make several o's and let them count them. (*Use letters for this purpose as fast as they are learned.*) Have them count the pupils in the class, the buttons on their jackets, the objects in this picture, etc.

As another exercise, use the pictures on pages 5 and 8. "All find the boys in the picture." "How many have found them?" (Hands raised to indicate.) "Count the boys (silently)." "Who can tell how many boys there are?". (Hands raised.) "Sarah, tell." "Mary, count them aloud." So proceed with other objects found in the pictures.

It will be serviceable as a class exercise to have the class repeat the numbers in concert, while you beat with the hand, bringing the right hand down into the open palm of the left at each count. Thus teacher and class count together. Class count *alone*, while the teacher beats. Class *beat and count*, while teacher only beats. Such exercises as these will be serviceable mainly in teaching the *names* and *succession* of the numbers; but this is no small part of the problem.

The *Numeral Frame* is very convenient in teaching counting.

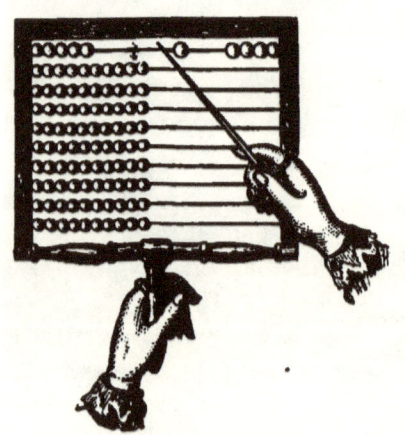

Teacher hold it up and slide out the balls one by one, as the class counts. Pupils take the pointer and slide out the balls and count, or use their fingers if they cannot handle the pointer. For other uses of this important instrument see pages 9, 13, 16, 18, 19, 20, 27, 31, etc.

Another exercise will consist of questions like the following: "How many eyes has each of you?" "How many feet?" "How many noses?" "How many fingers on one hand, without the thumb?" "How many with the thumb?" "How many on both hands, without the thumbs?" "How many with the thumbs?" etc. This exercise is more purely mental than the preceding, inasmuch as, in this, the pupils are expected to count the objects without touching them, or even looking at them. It should be extended to objects outside of the school-room (out of sight). Thus, "How many eyes has a cow?" "How many legs?"

" How many brothers have you?" "How many sisters?"
" How many brothers and sisters in all?" etc.

Seat Exercises.—While in the performance of other duties,
you may say to the "C" class, "The 'C' class may take out their
slates, *very carefully*." After a little while, when they are all
ready, say, "Each one make *three* marks on his slate." (To be
made thus, *///*. Perhaps it may be necessary to show them how,
by placing groups of marks in various positions on the blackboard, and instructing the pupils to make them in similar positions
on their slates.) Again, after a little time, say, " Each member of
the ' C' class make *four* marks," etc., etc.

The exercise may be varied by having dots made, instead of
marks. Better still, if they know how to make any letter, as *e*, by
having them make five *e's*, seven *e's*, etc.

Vary it again by having the pupils open to the pictures on pages
5 and 8, and tell them to make as many marks as they can find
dogs. As many as they can find kittens, etc.

Cautions and Suggestions.—Though this (learning to
count ten) is called *one* lesson, it will require several days, with
several class exercises each day, for pupils who know nothing of
counting at the outset, to master it. No class exercise should
occupy more than 5 or 10 minutes with this grade of pupils. The
seat exercises are quite as important as the *class exercises*. BE
SURE TO ALWAYS INSPECT THE WORK WHICH THEY ARE RE-
QUIRED TO DO ON THEIR SLATES. See that they do their best,
and do not merely scribble. Do not try to teach anything but
counting at this time. Defer teaching the characters (figures),
and how to make them, till another time. Do not distract their
attention with ideas of adding, counting backwards, or subtracting.
ONE THING AT A TIME. Nor need any *special* effort be made to
give the pupils the *idea* of number. If this idea is not innate,
they will get it from the above and kindred exercises. Generally,
one form (or at *most two*) of class exercise at a time is enough.
So also of a seat exercise. Short, single, clear, pointed, lively—
these are the characteristics of a good exercise.

LESSON II.

Purpose.—*To teach the Names and Meaning of the Figures,*

1 2 · 3 4 5 6 7 8 9 .

Also to review the preceding lesson in connection with this.

Method.—Class Exercise. With the class around the Counting Table and before the teacher's Blackboard, as represented in picture, page 1, make figure / on the board (make a simple, inclined, straight line, not *1*, nor any elaborate form). Then say, " Children, this means *one*." Pointing to it, ask, " What does this mean ? " " James, put out as many counters on the table as this means." " Jane, show me as many fingers as this means " (always pointing to the figure when the question is asked).

Again, make the figure 2 (in this simple form), and repeat the questioning as above. Thus, first telling them that it means *two*, ask them, " What does this mean? " " Hold up as many fingers as this means." " Mary, pick up as many counters as this means." So " question back" what you have told them. Do not be in a hurry. Put a great variety of questions, to the class, and to each member of the class. But let each question be directed to the one end of fixing in the mind the fact that the figure 2 means *two*. Do not call the figure by name, but point to it and say, " As many as *this* means."

Now put both 1 and 2 on the board. Pointing to 1, say, " How many does this mean ? " " Each show me as many hands as this means." " Each pick up as many counters as this means." In like manner point to 2, and question and exercise the class.

Proceed in like manner with 3. This will be enough for one exercise. (It may be too much for some classes.)

Seat Exercise.—When the time comes for this exercise, without turning aside from other duties, say, " The ' C ' class may take

out their slates." Put the figure 2 on the board. Say, "Each make as many marks (dots, e's, m's, a's, see preceding lesson) as this means." In two or three minutes, put the figure 1 on the board and proceed in the same manner. In like manner, after a short interval, put 3 on the board and direct as before.

Be sure to inspect the work after it is done.

A Second Class Exercise will teach the meaning of the figures 4, 5, and 6.

A Second Seat Exercise like the above will give them practice on it.

A Third Class Exercise and

A Third Seat Exercise, both similar to the above, will complete this part of the lesson.

A Fourth Class Exercise will teach them that *The names of the figures are the same as their meaning.*

A Fourth Seat Exercise may be given by telling them to make as many marks (dots, e's, m's, o's, see Lesson I) as figure 5 means. As figure 6, etc.

Cautions and Suggestions.—Do not let the exercises become monotonous. Strive to have the pupils come to them as to something which they enjoy. Requisitions which can be met by *doing something,* will be relished. Thus, "All clap hands 3 times." " All stamp as many times as this " (pointing to a figure). Counting in connection with gymnastic exercises, and the like, will enable you to accomplish several things at once, and aid in keeping up an interest.

The exercises of this lesson have been written out thus in detail to show how minute the subdivisions really need to be made. (See first *Caution* under the last lesson.) Generally, this subdivision will be left to the discretion of the teacher. Some classes will need shorter exercises than others—that is, will be able to take in fewer new ideas.

The *Numeral Frame* is very useful in teaching the meaning of figures. First let the *teacher* move the balls to show how many any given figure represents ; subsequently, let the *pupils* do the same.

LESSON III.

Purpose.—*To teach to make the Figures,*

1 2 3 4 5 6 7 8 9.

Also, to review, in connection with this, the two preceding lessons.

Method.—Take the figures of simplest form first. They will then be taken about in this order—1, 7, 4, 6, 9, 2, 3, 5, 8.

Let the first *Class Exercise* be to teach how to make 1, 7, and 4. "Who can tell me how to make the figure which means *one?*" Place the crayon point on the board, and lead the pupils to say "Mark right down." Make it too long—a foot long—and have them tell you, "It is too long." Make it crooked, and let them correct you. Incline it the wrong way, and let them tell what is the matter with it. Then ask each of them in turn to step to the board and make figure *one.* Call it sometimes "figure one," and sometimes "the figure which means *one.*" Ask frequently, "What is the name of this figure?" "What does it mean?" "Show me as many counters as it means?" "Clap hands as many times as it means," etc.

When a figure has been made on the board by a pupil, let the class notice wherein it is not right, and let those who notice defects try to make a better one. Thus promote a healthful ambition to do good work.

This is an outline of the procedure with every figure. Only a few farther suggestions need be made.

In teaching to make 7, first show the class *seven* objects. Let them count out seven counters, or count seven dots as you make them on the board. Then putting the crayon point on the board, have them tell you how to move it in order to make 7. They will say, "Towards the door, towards the clock," or give other similar directions. Lead them to say, "To the right," "To the left," instead. Marking a little way to the right, ask, "Which way now?" Lead them to notice that the stem is just figure 1.

Treat 4 in the same way. The general directions which the pupils should be led to give for making it, are, " Mark down—to the right—make a stem across the last line."

As you proceed to the more complicated figures, the importance of teaching the pupils to observe carefully each peculiarity of form increases. This will be best done by making the figure wrong in various ways, and telling them to correct, either by telling or by making a better one. Thus, make ✝ instead of 4. Make ⊔, or ⟂, or ⊔, etc.

Much ingenuity will be needed in helping the children to *child-like* descriptions of the forms of the various figures. But be sure and let them tell, in their own way, what the shape is, as far as they can. Their methods will often give you valuable hints.

Making 6, 9, and 2, will constitute a second *Class Exercise*. The stem of the 6 is bent (curved) to the left. Its back is bent. It turns up at the bottom into a little o at the right. It has a kind of mouth opening to the right, etc. (Children are fond of such conceits.)

Figure 9 is an o with a one for a stem. The stem is on the right. The o is at the top of the stem and on its left.

When the little ones attempt to make these characters for them-selves, either on the board or on their slates, you will often have to take hold of their hands and guide them. Manage the exercises so as to give each one a great deal of practice. The purpose is not to teach *how* to make the figures, but to teach the pupils *to make* them; and this can be accomplished only by much practice.

Have the slates brought to the *Counting Table* frequently, for a class exercise, and require the pupils to make the figures on their slates, as *you* make them on the board.

Figure 2 may be described as a hook, 2, with a foot to it, 2 —the foot running out to the right. The hook bends over to the left.

Making 3, 5, and 8, will constitute a third *Class Exercise*.

Figure 3 may be described as having two mouths opening to the left, or as two half o's one on top of the other. It has a very crooked back. Its back is at the right, etc.

Figure 5 may be described as having a short stem, a hook or

half o, or mouth opening to the left, and a handle or arm running out to the right from the bottom of the stem, etc.

The 8 is an S with a mark running up through it.

The true child's-teacher is fruitful in such comparisons.

Seat Exercises.—Mark off on the blackboard a representation of a slate, and put upon it a row of 1's. Let the pupils make a similar row on their slates. Then put a row of 7's on the picture of a slate, and let the pupils make them on their slates. Let them make several rows of 7's. Then rows of 4's, etc. This will constitute an exercise in *copying* figures, and will need to be continued for several days.

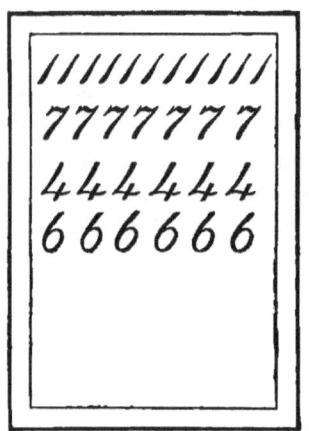

Another *Seat Exercise* may be given thus: Having no figures on the board, tell the class to make a row of 4's, of 2's, of 3's, etc. This is an exercise in remembering the forms of the figures so distinctly as to be able to make them. It will take several exercises.

Still another variety of *Seat Exercise* may be given by presenting objects, marks, dots, letters on the board, or holding up fingers, and telling them to make the figure which means so many.

Let them open their books to the pictures on pages 5, 8, and bid them make the figure which tells how many kittens there are. "Make it five times." "Make the figure which tells how many hats there are." "Make it seven times."

Caution.—Do not attempt to teach them to make anything but the simplest forms of figures. When they are older, and have learned to write, they will modify the style somewhat; but no one makes the more elaborate forms in common work on the slate or blackboard.

LESSON IV.

Purpose.—*To teach to Count from Ten to Nineteen.*

[A review of preceding lessons is *always* to be woven into a new lesson. Not *frequent*, but CONSTANT, reviews, is what we need.]

Method.—To accomplish this purpose will require several days, with several exercises each day. The first class exercise will *begin* the work of teaching to count from ten to fifteen. Move out the counters one by one while the pupils count till you have moved out *ten*. Now, putting another with these, ask, "How many have we now?" Of course, none are expected to know. But arouse their curiosity, create a desire to know, and *then* tell them, "eleven" (not "*leven*"). So proceed to fifteen. Go slowly. Repeat each step several times.

The *Numeral Frame* is exceedingly convenient for this purpose. Sliding all the balls to one side of the frame, move all on the top wire to the other side. Call attention to the fact that there are *ten* in this row. Then, moving out one on the next wire, then two, then three, etc., and requiring the pupils to observe, as you do so, the composition of these numbers is clearly illustrated to the eye.

Let the class name in concert the numbers in order from ten to fifteen, while you beat with the hand. Class beat and count. Call on individuals to count while you beat.

Again, use the counters. Have individuals move them out and count, while the class watches to see if it is done right.

Use the pictures at the close of this lesson. Let the pupils open their books to it. "How many can find the nest?" "How many eggs are there in it?" "How many by the side of the nest?" "How many in all?" "*Ten* and *one* are called what?" "How many fishes *up* in the water in the vase?" "How many on the bottom?" "*Ten* and *four* are called what?" etc., etc.

When they can count to fifteen, call their attention to the fact that we count from twelve by saying "*thir*-teen," which means *three* and ten (or *three*-teen); "*four*-teen," or four-and-ten, the teen meaning *and-ten*; *fif*-teen, or *five*-and-ten, etc. Then by means of the counters show them that *thir*-teen is *three*-and-ten, *four*-teen is *four*-and-*ten*, etc. Do not go beyond fifteen until this idea is clearly perceived by all. Use the subsequent numbers, sixteen to nineteen, as tests. Thus, if they have comprehended the idea, they will be able to tell you what six-and-ten is called, what seven-and-ten, etc. To test the members of the class separately use the pictures below. Repeat the numbers in concert.

You will very naturally speak of these numbers as "the teens." There is no impropriety in it, and it will help to fix them, *as a class*, in the pupils' minds.

Seat Exercises.—"Each make eleven figure 2's." After awhile, "Each make twelve dots." Again, having given time for the former, "Each make thirteen 1's," etc. These directions are to be given without interfering with your other duties, in a quiet manner and low tone, addressed to the "C" class in their seats.

Be sure to inspect their work.

LESSON V.

Purpose.—*To teach to Write the Numbers from Ten to Nineteen.*

| **Method.**—**Class Exercise.**—Make 0 on the board, telling the class its name, *Cipher,* or *Zero.* Show them the difference between the form of 0 and o.

Having taught the *form* and *name* of this character (you need say nothing of its meaning, or use), let the class count while you make ten marks on the board, thus, / / / / / / / / / /. Write 10 over the marks, and tell them that these figures, so written, mean *ten.* Then, pointing to the figures, ask, " What do 1 and 0 mean when written so ? "   " How is ten written in figures ? "   Write the 0 *over* the 1, *under* it, on the left of it, and ask in each case, " Is this ten ? "   " What figures *do* mean ten ? "   " How written ? " " John, step to the board and write the figures which mean ten." " Mary, write them."   " James," etc.   Proceed in a similar manner to teach that 11 means eleven ; 12, twelve ; 13, thirteen.

When they have gone thus far, call attention to the fact that you are just writing the figures 1, 2, 3, 4. 5, etc., in order, after 1. Thus, eleven is written by putting 1 after 1, twelve by putting 2 after 1, thirteen by putting 3 after 1, etc.   Illustrate this by writing on the blackboard, thus :

0 1 2 3 4 5 6 7 8 9 10 11 12 13, etc.

So illustrate and impress the principle, that the pupils can go on from 13 to 19.   Such questions as this will aid in this work : " What is the first figure in writing each of the *teens ?* "   " Well, then, what shall I write first if I wish to write seventeen ?   Get the answer from each pupil.   " What number am I to write ? " (*Seven*-teen.)   " What teen ? "   (*Seven.*)   " What, then, shall I write after the 1 ? "

To enliven the class, call on Mary to take up a handful of counters and lay them in a pile. Let Jane count them, and Henry
write the number on the board. Sarah add a handful to Mary's.
James count *this* pile. John write the number, etc.

Still another useful exercise will be secured by writing the
numbers from 1 to 19 on the board, thus:

$$0 \quad 1 \quad 2 \quad 3 \quad 4 \quad 5 \quad 6 \quad 7 \quad 8 \quad 9$$

$$10 \quad 11 \quad 12 \quad 13 \quad 14 \quad 15 \quad 16 \quad 17 \quad 18 \quad 19 \,,$$

and then, as you point to them, promiscuously, let the pupils tell
what they mean. Make this also an individual exercise. Name
different numbers and call upon pupils to point to the figures on
the board.

For another exercise, the pupils may bring their slates to the
table and write the numbers as you dictate them.

Seat Exercises.—Dictate numbers to be written on the slates.
Write the numbers from 1 to 19 on the board and have them
copied. Let the class stand at the long board and write the numbers you name. (Such an exercise should not last more than two
or three minutes, and should not interrupt other duties. Thus
have the "C" class take places at the long board while the "B"
or "A" class is coming out for class exercise. They should re-
turn, quietly, at a signal, while other exercises are going forward.)

Make dots, marks, or letters, on the board, and bid them make
the figure or figures which mean "so many."

The pictures on page 14 may be used for this purpose, by telling
the pupils to make the figures which tell how many eggs there
are, when counted together; how many ants, books, mice, etc.

If the pupils in their seats seem restless, have them stand and
count in concert. They can count twenty, three or four times in a
minute. They may have such an exercise half a dozen times a
day.

LESSON VI.

Purpose.—*To teach the Names and Meaning of the Decades, Twenty, Thirty, Forty, etc., to One Hundred.*

Method.—Class Exercises.—"Children, if you count the thumbs as fingers, how many fingers has James on both hands?" "James, place your hands side by side on the table" (as in the margin). "Now, how many fingers (including thumbs) are there?" "How many *tens?*" "Now, James, place your hands close together, and Henry, put yours down by the side of James's." (See margin.) "Now, how many *tens* are there?" "How many tens has James?" "How many has Henry?" "How many have both together?" (Two.) "What do we call *two tens?*" (Twenty.)

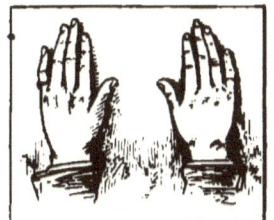

The accompanying cut may be used for this purpose. All having their books open to it, ask, "How many rails in this fence?" "How many birds on one of the top rails?" "Count the birds on

each of the rails?" Lead them to notice that there are *ten* birds on each rail. "Now, on two rails there are how many tens?" "What are two tens called?" etc.

Proceed in like manner to teach what is meant by *Thirty, Forty, Fifty*, etc., to *Ninety*. Repeat these names in concert.

Call attention to the prefixes *Twen* (meaning *two*), *Thir* (three), *For* (four), *Fif* (five), *Six*, etc., and to the *ty* as meaning *tens*. Thus, *Six-ty* means *six*-tens, *Seven-ty* means *seven*-tens, etc.

Have concert exercises like this : " *Twenty*—means *two* tens ;" " *Thirty*—means *three* tens ;" " *Forty*—means *four* tens ;" etc.

Again, let part of the class name the decades, and the other part tell what they mean. Thus, *First Part*, in concert, " Twenty " —*Second Part*, "Means two tens ;" *First Part*, " Thirty "—*Second Part*, " Means three tens," etc.

Groups of dots, or marks on the board, ten in a group, may be used to show the meaning of these terms.

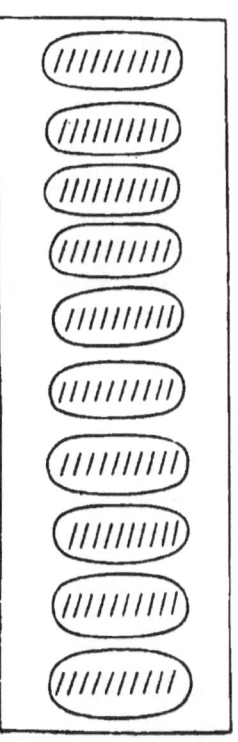

Seat Exercises.—"Each make ten marks on your slate and draw a line around them, as I do on the board." " Make another ten right under these, and draw a line around them, as I do." So proceed till they have ten groups of tens. After they have made two or three groups, the single word " Another," spoken by the teacher, will be sufficient direction to keep them at work. The *Numeral Frame* may be used very conveniently for this purpose.

These marks on their slates may be made to afford an excellent *Class Exercise*. Thus, have the pupils bring their slates to the *Counting Table*. "John, rub out ten of Henry's marks." " Henry, rub out *thirty* of your marks." " Mary, rub out *twenty* of Henry's marks." " Henry, how many marks have you now ?" (This is not for the purpose of teaching Subtraction, but simply to teach what twenty, thirty, forty, etc., mean. Do not ask them how many twenty from sixty leaves, or any such questions. They are out of place here. *Stick to the single purpose.*)

Another *Seat Exercise* may be obtained by having the ten groups of tens made as before, and then telling them to draw a line around twenty, thirty, forty, etc.

LESSON VII.

Purpose.—*To teach to Write the Decades, as,*

/0, 20, 30, 40, 50, 60, 70, 80, 90.

Method.—Class Exercises.—Show how ten, twenty, and thirty are written, and then call attention to the fact that each has a 0 (zero) at the right—that ten (*one* ten) has 1 at the left ; twenty (*two* tens) has 2 at the left ; and that thirty (*three* tens) has 3 at the left. Illustrate this by writing the decades in a column on the blackboard as far as thirty, and lead the pupils to complete the work on their slates.

```
 0
10
20
30
etc.
```

Seat Exercises.—Direct the pupils to make the figures which mean twenty, thirty, forty, etc. Make groups of ten marks each, on the board, as in the preceding cut, and making a mark around two, three, four, or any number of them, say, "Make the figures which mean so many," etc. Slip three tens of the balls on the *Numeral Frame* to one side, and bid them make the figures which mean so many. Then forty, then fifty, etc.

Use the picture on page 17. "Make the figures which tell how many birds there are on two rails." "On three," etc.

LESSON VIII.

Purpose.— *To teach to Count through the Decades.*

Method.—Class Exercises. — Have the pupils count out twenty (two tens) of counters, and place them by themselves. Put another with them, and lead the pupils to tell you that there are " Twenty and one counters." Tell them that we call it " *Twenty-one*," instead of " twenty and one." Put another counter with these, and in like manner lead them to tell you that there are "Twenty and two counters." Tell them, "We call so many, *Twenty-two*, instead of twenty and two." Thus lead them to count *through* the *twenties*. Then through the thirties, etc. They should be able to go on of themselves after having been led through two or three decades. The *Numeral Frame* is well adapted to this purpose.

There will be needed much drill in repeating in concert the names of the decades, and in counting *through* them. Many class exercises will need to be devoted mainly to simple drill in oral counting from *one* to one hundred.

Arouse the ambition of the pupils to be able to count one hundred! Test them individually. Give them certificates that they can count to one hundred, when they can do it well. This is one of the greatest mathematical achievements they will ever make! Vary the counting exercise by having one count awhile, then another go on a little further, then another, etc. Also by having them count around the class. Thus, beginning at one end, the first pupil says, " one ; " the second, " two ; " the third, " three," etc., to one hundred. This should be kept up for a long time, till all are perfectly familiar with the order of the numbers.

Seat Exercises.—" Each make twenty marks." (Two groups of ten each.) " Make enough more so that you will have twenty-three in all." " Enough more to make twenty-five," etc. So of thirty, forty, etc.

"Make three lines of marks clear across your slates" (show them how on the board). "Count them, and tell me how many you have."

Use the picture below. "How many birds?" "Lambs?" etc. Show from the picture that 24 is 2 tens and 4, etc.

LESSON IX.

Purpose.—*To teach to Write in Figures through the Decades.*

Method.—Class Exercises.—Let one pupil count out ten counters and put them in one pile on the counting-table, and have another count out ten more and place them in a pile near the former. "How many *tens* have we here?" "What do we call *two tens?*" "Tell me how to write twenty in figures, on the board." "John, what shall I write first?" "Jane, what next?"

Now put the two piles of tens together and lead the pupils to say that there are twenty in the pile. Then put a single counter near them, and ask, " How many have we here in all ? " " Twenty and one are called how many ? " " What comes next after twenty ? "

Then calling attention to the 20 on the board, tell them that the 2, in the second place from the right, means so many (two) *tens*, and that whatever stands in the first place at the right means *so many more.** Rub out the 0 and put 1 in its place, thus, 21. " How many does the 2 mean ? " " Why does the two mean two-tens, or twenty ? " (Because it stands in the second place from the right.) " What does this mean ? " (pointing to the 1.) " Twenty and one we call what ? " Then write 25, and proceed in like manner. Then 27, etc. (Do not use 22 until the *idea* is fixed. The two 2's may trouble them.) Practice upon this until each pupil clearly perceives the principle, and has it fixed in the mind. You need not confine yourself to the twenties in doing this, nor to the numbers in order ; it will be better that you should not. Thus, write 57. " What does the 5 mean ? " (Five tens, or fifty.) " What this ? " (pointing to the 7.) " What do we call fifty and seven ? " " Then what does this mean ? " (pointing to the 57.) Then take 43, 65, 72, or any of the numbers.

Whether the fact that in 10, 20, 30, 40, etc., the 0 *is good for nothing but to keep the place,* should be taught or not, will depend upon the ability of the class. I think it can usually be done to advantage. Thus, write 5. " What does this mean ? " Put 0 by the right side of it. " What does the 5 mean *now?* " Show them that the 0 helps the 5 to mean fifty, by showing that it is in the second place. *It does nothing else.* Now write 5 again. " What does this mean ? " (Five.) Put 3 at its right ; thus, 53. " What does the 5 mean now ? " " What helps it to mean fifty ? " " Does the 3 do anything but help the 5 ? " *It* does something by itself. It means *three.* In this manner teach that 0 only serves to keep the place, and show that the other figure is in the second place, while any other figure will keep the place and do some-thing else.

---

* This form of speech is given as that which the teacher will naturally use, and which she will make clear by her manner.

If these *principles* have been properly taught, the pupils can now tell you how to write any number in any decade. Test them on this point, as a means of determining when your work is well done. Thus, " I want to write forty-five." " How many ? " (Class say "forty-five.") " How many tens? " " How many more ? " " What shall I write to mean the tens ? " " How many more do I want to write ? " " Where shall I write the 4 ? " etc. This drill must be kept up exercise after exercise, day after day, till all can write any number up to 99, readily.

Illustrate the writing of numbers through the decades by writing them on the board in this form :

```
0   1   2   3   4   5   6   7   8   9
10  11  12  13  14  15  16  17  18  19
20  21  22  23  24  25  26  27  28  29
30  31  32  33  34, etc.
```

Let the pupils copy this arrangement and carry it forward to 99. Drill them in reading the numbers across the page, and also down the columns.

The single statement that 100 means ten tens, or one hundred, will be enough on this point, if it is illustrated and dwelt upon till the pupils know it.

Seat Exercises.—These will be easily devised. " Each write twenty-six." " Each write fifty-seven." (They will understand what is meant without saying " the figures which mean," etc. although this full form of expression should be kept up till the thought conveyed is fixed in mind.)

" Write fifty-eight three times." " Write sixty-three." " Write seventy-one under the sixty-three," etc., etc.

Use the pictures, having the pupils count the various objects, and write the number in figures. Thus, on page 14, " Count the fishes, the ants, and the lilies of the valley, and write the number." Such a demand as this will require several minutes for its execution, and you should by no means fail to examine the results. Do not forget that it is a great work for the little ones.

Now have an exercise in "finding the page" in their books. "All turn to page thirty-seven." Require them to hold up their books so that you can see from your desk that they have found the right page. "All find page fifty-three." "Show it to me," etc., etc. Drill on this till all can turn quickly to any page you may name.

LESSON X.

Purpose. — *To teach the Ordinals, or how to Number.*

Method.—This may be done by having the class number around. Thus, one at one end says "first;" the next, "second;" the next, "third," etc.

It will not be best to make this an entire exercise, but spend a little time upon it, and the rest of the time on review exercises in counting, writing, and recognizing numbers.

Make figures thus on the board, *beginning at the right,*

$$9 \quad 8 \quad 2 \quad 3 \quad 4 \quad 6 \quad 7 \quad 5 \quad 4$$

"Which figure did I make *first?*" "Which, *second?*" "Which, *third,*" etc.

Let it be understood that you expect them to number from the right, and then ask, "What is the *fourth* figure?" "What the *seventh?*" etc.

Seat Exercise.—"Find the *twenty-first* page." "The seventeenth," etc.

This lesson will require several days, and but few of the ordinals should be attempted at a time. Perhaps for the first exercise from "first" to "tenth."

It is not imperative that the numbering should be carried to one hundredth, at present; perhaps to thirtieth, will be far enough before going on to other lessons. But if they see clearly the principle they may be able to go to one hundredth without difficulty.

SECTION II.
ADDITION.

LESSON I.

Class Exercises.

Purpose.—*To teach how to find out the sum of any two numbers between one and nine.*

* This lesson is not, strictly speaking, a lesson in Addition; —it is a lesson in Counting, and is preparatory to Addition. Addition and counting are not the same thing. The arithmetical process which we call Addition is a method of finding the sum of numbers by means of a knowledge of the sums of the digits two and two, *i. e.*, by means of a knowledge of the *Addition Table*. Hence, as preparatory to Addition, the pupil needs—

---

* The paragraphs in small type are exclusively for the *Teacher's* use. Those in the larger type are for the children to study at their seats.

1. *To learn to make the Addition Table;* and

2. *To commit this table to memory.*

Having done this he is ready to learn Addition proper.

The *only* way in which the pupil can find out, in the first instance, what is the sum of any two numbers, as 4 and 5, is by taking one number and counting on the other. *But this is not addition.* When the pupil has in this way learned the meaning of the Addition Table, and can make it readily, he is prepared to commit it to memory. This is the second step : and not until the table is thoroughly learned, is the pupil prepared to enter upon Addition. It is just at this point that all the difficulty in teaching Addition occurs. The pupil is allowed to attempt adding before he is familiar with this table ; hence he necessarily falls into the habit of counting. But if he is not allowed to enter upon Addition proper (Lesson III.) until he can tell *with perfect ease* the sum of any two digits, *at sight*, there will be no trouble arising from a propensity to count. This propensity arises solely from an imperfect knowledge of the Addition Table.

We now proceed to exhibit in detail some methods of conducting class exercises for the purpose designated at the head of this lesson.

Have the class count while you place *four* counters in one pile, and three counters in another pile near the first. " How many are there in this pile ? " (pointing to the *four*.) " How many in this ? " (pointing to the *three*.) " Now, who can tell how many there are in both piles ? " Of course it is not expected that any one can. *But arouse the desire to find out.* Then show them how, by beginning with the four, and counting on the three, they can find out how many there are in both piles. Thus, ask, " How many are there here ? " (pointing to the four.) Move one of the three up to the four. " How many now ? " Move up another. " How many now ? " The other. " How many now ? " " Now we have put the three with the four." " How many are four and three together ? "

Again, place 5 in one pile and 6 in another, and teach them *how to find out*, by counting, how many 5 and 6 put together make.

Also lead them to determine how many 5 and 6 make by placing the counters in each collection so that they can be counted without being moved.

The *Numeral Frame* may also be used for our present purpose. Thus, holding it up before the class, let the pupils count out 5 balls as you move them to one side on the upper wire. Then count out 4 on the second wire, moving them under the five. "How many balls have we here?" (pointing to the 5.) "How many have we here?" (pointing to the 4.) "How many in all?" "How many are 5 and 4?"

Again, propose the question, "How many are 5 and 3?" and let the pupils work out the answer by moving the balls. So, also, address such questions as the following to *individuals*, and let them find out the answers by moving the balls: "How many are 4 and 2?" "How many are 7 and 3?" etc.

Counting by 2's, 3's, 4's, etc., is a very useful exercise for many purposes, especially as it furnishes such a variety of systematic exercises in a convenient form for drill. But let the exercise be restricted to the single purpose had in view at the time : our present object is *to teach to make the Addition Table.* For this purpose, as well as for the purpose of fixing the table in memory, it is not legitimate to carry this form of counting beyond those steps which require the combination of *single digits ;* for example, in counting by 2's we shall have these combinations, 2, 4, 6, 8, 10 ; 1, 3, 5, 7, 9, 11 ; and *no more.* Counting by 3's we shall have 3, 6, 9, 12 ; 1, 4, 7, 10 ; 2, 5, 8, 11 ; and *no more.* After the pupils comprehend the *order*, these combinations can be assigned as *seat exercises ;* for example, tell them to write on their slates the numbers as they count by 4, first beginning with 1. These results will be written thus :

1, 5, 9, 13.

Again count by 4's beginning with 2, and write the results. These will be 2, 6, 10.

The several exercises thus outlined will be distributed through a number of days, *only one being used at a time*, and this repeated till it is familiar before passing to another.

Do not attempt, at this time, to have them *remember* how many any combination makes. The present purpose is merely to learn *how to find out* what 4 and 3, 5 and 6, etc., make. In this first exercise do not take either the smallest or the largest numbers.

Give them sufficient practice so that they can study the following seat exercise. Write the figures 4, 8, 5, 2, 1 on the board, with the corresponding words printed under them, thus,

4   8   5   2   1
Four.   Three.   Five.   Two.   One.

If the pupils do not know all these words, they should be taught them, or at least be shown how to find out what they are by looking at the board.

Seat Exercise.

1. How many birds are on the box?
2. How many birds are in the tree?
3. How many birds in all?
4. How many birds are four birds and three birds?

5. How many birds are on the barn?
6. How many birds are flying to the barn?
7. How many birds in all?
8. How many birds are five birds and three birds?

9. How many cats are on the table?

10. How many cats are on the floor by the table?

11. How many cats in all?

12. How many cats are three cats and three cats?

13. How many boys are at play under the tree?

14. How many girls are at play under the tree?

15. How many boys and girls in all?

16. How many are 2 boys and 3 girls?

17. How many caps are hung up?

18. How many caps are on the floor?

19. How many caps are there in all?

20. How many caps are 4 caps and 2 caps?

Recitation and Class Exercise.

We are now to have our first *Recitation*.

With books in hand, let the pupils read the questions in the preceding *Seat Exercise*, and give the answers. They are not expected to have learned the combinations, as how many 4 and 3 make, but only *how to find out by counting*, as above explained. Give time to do this in the recitation. To give variety, let one read a question and another answer it. But, *do not go round in order*. Say, "*All* read the first question, silently." Give time. "Jane, read it aloud." "John, answer it." Thus, letting no one know who is to read, or who is to answer, keep all in readiness.

When they have had a fair opportunity to show that they studied the lesson well, give them a new exercise.

Give them sufficient practice to enable them to find out the answers to the following, while in their seats, *and write the answers in order on their slates, while in their seats.*

Seat Exercise.

1. How many are 4 and 3 ?	9. How many are 7 and 6 ?
2. How many are 3 and 6 ?	10. How many are 8 and 5 ?
3. How many are 2 and 3 ?	11. How many are 6 and 1 ?
4. How many are 5 and 1 ?	12. How many are 9 and 2 ?
5. How many are 4 and 1 ?	13. How many are 8 and 3 ?
6. How many are 3 and 1 ?	14. How many are 7 and 9 ?
7. How many are 2 and 7 ?	15. How many are 6 and 7 ?
8. How many are 6 and 8 ?	16. How many are 7 and 8 ?

Recitation and Class Exercise.

First, the pupils having brought their books, and their slates with the answers in order on them, read the questions to them, and let them give the answers as they have them on their slates. Let several pupils give the answer which they have to each problem, as it is called for. If they do not agree, have the class find out which is right. It may take several exercises to get them all to write their answers in good order on their slates ; but the effort should be repeated and persisted in until they do it.

Second, having gone through with all the questions, and given all the pupils a full opportunity to exhibit their work, give a *new exercise.* Let this be to teach them how to use marks like **/ / / /**, and the balls on the Numeral Frame, for the purpose of finding out what the sum of two numbers is. (See page 27.)

The following exercise is to be studied by the pupils in their seats, and the answers written in order on their slates in the same manner as the last.

Seat Exercise.

1. How many are 5 and 9 ?
2. How many are 6 and 1 ?
3. How many are 7 and 4 ?
4. How many are 4 and 9 ?
5. How many are 5 and 2 ?
6. How many are 8 and 9 ?
7. How many are 3 and 7 ?

8. How many are 2 and 9 ?
9. How many are 1 and 1 ?
10. How many are 2 and 1 ?
11. How many are 5 and 6 ?
12. How many are 8 and 7 ?
13. How many are 3 and 9 ?
14. How many are 1 and 8 ?

Recitation and Class Exercise.

The recitation will be similar to the last, the design being to satisfy yourself that the pupils have done the work well, which was assigned them to do in their seats, and to make them feel that you notice and appreciate their efforts.

For a *New Class Exercise* teach them that + means the same as "and," and =, the same as "make," or "are." Write on the board $4 + 5 = 9$, $6 + 4 = 10$, and similar expressions, and teach them to read them "4 and 5 are 9," "6 and 4 are 10," etc. Then give them what instruction they may need to enable them to copy the next seat exercises upon their slates, and to determine the answers by counting, with or without objects, and to fill out the expressions.

Seat Exercise.

$3 + 5 =$	$8 + 9 =$	$7 + 4 =$	$3 + 7 =$
$2 + 3 =$	$7 + 6 =$	$6 + 9 =$	$0 + 0 =$
$6 + 4 =$	$3 + 1 =$	$8 + 8 =$	$2 + 0 =$
$7 + 1 =$	$4 + 1 =$	$6 + 6 =$	$3 + 5 =$
$5 + 6 =$	$8 + 3 =$	$4 + 0 =$	$9 + 4 =$

Recitation and Class Exercise.

Examine the pupils' slates to see that the work is done neatly. Question them thus: "6 and 4 are how many?" "7 and 6?" etc. When a question is asked, have *all* look up the answer on their slates, and then call on some one to answer, allowing the others to correct the reply, if wrong.

For a new exercise, show them *how to make the Addition Table*, as indicated in the following exercise. Their slates are to be ruled, and the table copied and filled out.

Let it be borne in mind that it is *ability to find out by counting*, what the sum of any two numbers each expressed by a single figure is, that we are seeking to secure. We are not now requiring the pupils to *memorize*, but to *make* the *Addition Table*.

Seat Exercise.

$1 + 1 =$	$1 + 2 =$	$1 + 3 =$
$2 + 1 =$	$2 + 2 =$	$2 + 3 =$
$3 + 1 =$	$3 + 2 =$	$3 + 3 =$
$4 + 1 =$	$4 + 2 =$	
$5 + 1 =$	$5 + 2 =$	
$6 + 1 =$	$6 + 2 =$	
$7 + 1 =$	$7 + 2 =$	
$8 + 1 =$	$8 + 2 =$	
$9 + 1 =$	$9 + 2 =$	
$1 + 4 =$	$1 + 5 =$	$1 + 6 =$
$2 + 4 =$	$2 + 5 =$	$2 + 6 =$
$3 + 4 =$	3	
4		
5		
6		
7		
8		
9		
$1 + 7 =$	$1 + 8 =$	$1 + 9 =$

Encourage the pupils to make a neat *Addition Table* on paper, at home, and bring you a copy *to keep.*

To make, and thoroughly to memorize this table is a great achievement. Let the pupils know that their teacher thinks they are doing " great things."

Recitation and Class Exercise.

Making the above table will afford three or more seat exercises. Each part should be made several times over, until all can be made with ease. The class exercises will be similar to that suggested last. Remember that the present purpose is *to learn how to find out* what these combinations are. The next lesson will be devoted to fixing them in memory.

A good class exercise can be obtained by writing a series of combinations on the board, thus :

$$3 + 2 =$$
$$2 + 4 =$$
$$1 + 5 =$$
$$6 + 3 =$$
$$8 + 5 =$$

etc. ;

and as you point to any combination, let the pupils raise their hands as soon as they can tell what it makes. Then call on individuals to answer.

Seat Exercise.

1. A boy has three apples and a girl gives him five more. How many has he then?

2. Frank has 2 tops and George has 3. How many have they both?

3. There are 3 birds on one tree and 8 birds on another. How many birds are there on both trees?

4. Ann has seven flowers and George gives her six more. How many has she then?

5. There are 5 books on the chair and 8 on the table. How many books are there in all?

6. There are 4 chickens in the barn and 7 in the yard. How many chickens are there in all?

7. There are 5 eggs in one nest and 7 in the other. How many eggs in both nests?

8. If the black cat has four kittens and the white cat has six, how many kittens have both?

9. How many letters are there in the word ground? How many in the word white? How many in both words? How many are six and five?

10. How many letters are there in the word teacher? How many in the word boy? How many in both words? How many are 7 and 3?

[See first Recitation and Class Exercise for method in this case.]

LESSON II.

Purpose.—*To fix the Addition Table in the memory, so that the pupil can tell the sum of any two numbers between 1 and 9 with readiness.*

Class Exercise.

Show them, by the use of the counters or other objects, that 3 and 2 are the same as 2 and 3; 5 and 4, as 4 and 5, etc.

"If you have three nuts in your left hand and 5 in your right,

how many nuts have you in all?" "If you change and take 5 in
your left hand and 3 in your right, how many have you then?"
"3 and 5 are the same as what?"

Be sure that this is understood and fixed in mind. It diminishes
the work of learning the addition table one-half.

When this principle is well learned, drill them in concert on the
1's of the Addition Table. Thus, have them all say, "1 and 1 are
2," "2 and 1 are 3," "3 and 1 are 4," etc.

Again, write on the board such examples as these :

1	1	1	1	1	1	6	1	7	1	8
3	5	4	1	2	6	1	7	1	8	1

Then, as you point to the example, ask, "How many are 3 and
1?" Tell them, "We will write the answer right under the
line." Having done it, proceed in like manner with the rest.

Show them how to perform the following exercise, by copying
it on their slates and writing in the answers. Each of the nine
following exercises are to be thus copied and filled out, and the
FIRST COLUMN IN EACH THOROUGHLY MEMORIZED.

Seat Exercise.

$1 + 1 =$	$2 + 1 =$	$6 + 1 =$	$1 + \ \ = 8$
$2 + 1 =$	$1 + 2 =$	$1 + 6 =$	$1 + \ \ = 9$
$3 + 1 =$	$3 + 1 =$	$7 + 1 =$	$1 + \ \ = 7$
$4 + 1 =$	$1 + 3 =$	$1 + 7 =$	$1 + \ \ = 6$
$5 + 1 =$	$4 + 1 =$	$8 + 1 =$	$1 + \ \ = 4$
$6 + 1 =$	$1 + 4 =$	$1 + 8 =$	$1 + \ \ = 3$
$7 + 1 =$	$5 + 1 =$	$9 + 1 =$	$1 + \ \ = 10$
$8 + 1 =$	$6 + 1 =$	$1 + 9 =$	$1 + \ \ = 2$
$9 + 1 =$	$1 + 0 =$	$0 + 1 =$	$1 + \ \ = 5$

1. George has three pigs and Frank has one pig. How
many pigs have both?

2. Mary has 1 chicken and Jane has 7 chickens. How many chickens have both?

3. One and 3 are how many?

4. One and what make four?

5. Three and what make four?

6. Five and one are how many?

7. Five and what make six?

8. One and what make six?

9. Mary has 4 flowers. How many more must she get to have five?

10. George has 1 top. How many more must he get to have 4?

| 1 | 1 | 2 | 1 | 3 | 1 | 4 | 1 | 5 | 1 | 6 | 1 | 7 | 1 | 8 | 1 | 9 |
| 1 | 2 | 1 | 3 | 1 | 4 | 1 | 5 | 1 | 6 | 1 | 7 | 1 | 8 | 1 | 9 | 1 |

Recitation and Class Exercise.

First, EXAMINE THE PUPILS' WORK.

Have the pupils repeat the "o ᴇs column" (the left-hand one) down and up, by having the first pupil say, "One and one are two ;" the second, " Two and one are three ;" and so on around the class. Then say it backwards in the same manner.

Then vary the exercise by having one say, " Nine and one are ten ;" and the next, "One and nine are ten ;" etc.

Dictate the second and third columns, and have the pupils answer. Thus, say, "1 and 2 are ——?" and when all have thought, name some one to answer. So of the others.

Again, have the pupils read the fourth column, supplying the answers as they read.

Give the necessary instruction to enable the pupils to prepare the next lesson. This and the subsequent exercises of this lesson are to be copied on the slates and treated as the last. The same general plan of recitation will be pursued in each of the eight following exercises.

In talking with the class about these exercises as they come to them, show them that 1 + 2, in the 2's column, 1 + 3, and 2 + 3, in the 3's column, etc., have been previously learned.

The teacher needs to bear in mind that ability to *add by sight* is quite as important as ability to add by *sound*, and adapt the drill exercises to the fact. Again, a fundamental purpose in these exercises is *to teach the component parts of the numbers from 2 to 18, so that they will be instantly recognized.*

Seat Exercise.

	$1 + 2 =$	$6 + 2 =$	$2 + \quad = 4$
$2 + 2 =$	$2 + 1 =$	$2 + 6 =$	$2 + \quad = 6$
$3 + 2 =$	$2 + 2 =$	$7 + 2 =$	$2 + \quad = 8$
$4 + 2 =$	$3 + 2 =$	$2 + 7 =$	$2 + \quad = 5$
$5 + 2 =$	$2 + 3 =$	$8 + 2 =$	$2 + \quad = 10$
$6 + 2 =$	$4 + 2 =$	$2 + 8 =$	$2 + \quad = 10$
$7 + 2 =$	$2 + 4 =$	$9 + 2 =$	$2 + \quad = 3$
$8 + 2 =$	$5 + 2 =$	$2 + 9 =$	$2 + \quad = 7$
$9 + 2 =$	$2 + 5 =$	$0 + 2 =$	$2 + \quad = 9$

1. Ann has 8 flowers and George has 2. How many have both ? If Ann has 2 and George 8, how many have both ?

2. Frank has 2 hens and George has 6. How many have both ? If Frank has 6 and George 2, how many have both ?

3. Two white pigs and four black pigs are how many ? Four white pigs and 2 black ones are how many ?

4. Two and three are how many ?

5. Two and what make five ?

6. Three and what make five ?

7. Eight and two make how many ?

8. Eight and what make 10 ?

9. Two and what make 10 ?

10. There are 3 eggs in the nest. How many more must the hen lay to make 5 ?

11. If James has learned 6 words, how many more must he learn to know 8 ?

1	2	2	2	3	2	4	2	5	2	6	2	7	2	8	2	9
2	1	2	3	2	4	2	5	2	6	2	7	2	8	2	9	2

Seat Exercise.

	1 + 3 =	6 + 3 =	3 + = 8
	3 + 1 =	3 + 6 =	3 + = 9
3 + 3 =	2 + 3 =	7 + 3 =	3 + = 7
4 + 3 =	3 + 2 =	3 + 7 =	3 + =11
5 + 3 =	3 + 3 =	8 + 3 =	3 + =10
6 + 3 =	4 + 3 =	3 + 8 =	3 + =12
7 + 3 =	3 + 4 =	9 + 3 =	3 + = 4
8 + 3 =	5 + 3 =	3 + 9 =	3 + = 5
9 + 3 =	3 + 5 =	0 + 3 =	3 + = 6

1. George has 7 books and Mary has 3. How many have both?

2. There are five lambs in one field and three in another. How many are there in both fields?

3. If there are 9 cows in one field and 3 cows in another, how many are there in both fields?

4. Eight eggs in the nest and three in your hand, make how many eggs?

5. Four and three are how many?

6. Three and what make seven?

7. Four and what make seven?

8. Five and four are how many?

9. Five and what make nine?

10. Four and what make nine?

11. Four and eight are how many?

Seat Exercise.

	1 + 4 =	6 + 4 =	4 + = 6
	4 + 1 =	4 + 6 =	4 + = 7
	2 + 4 =	7 + 4 =	4 + = 9
4 + 4 =	4 + 2 =	4 + 7 =	4 + = 12
5 + 4 =	3 + 4 =	8 + 4 =	4 + = 5
6 + 4 =	4 + 3 =	4 + 8 =	4 + = 10
7 + 4 =	4 + 4 =	9 + 4 =	4 + = 8
8 + 4 =	5 + 4 =	4 + 9 =	4 + = 13
9 + 4 =	4 + 5 =	0 + 4 =	4 + = 11

1. Four girls and five girls are how many girls?
2. Five boys and four boys are how many boys?
3. Nine pigs and four pigs are how many?
4. There are 7 sheep in one field and 4 in another. How many are there in both?
5. How many letters are there in the word George? How many in the word read? How many letters in both words?
6. George has 4 tops. How many more must he get to have 6? How many to have 9?
7. Mary has 3 flowers. How many more must she get to have 7?
8. Jane has 7 chickens. How many more must she get to have 11?
9. One day the hen's nest had 9 eggs in it. On another day it had 13. How many new eggs had been laid in it?

1	4	2	4	3	4	4	5	4	6	4	7	4	8	4	9	4
4	1	4	2	4	3	4	4	5	4	6	4	7	4	8	4	9

Seat Exercise.

	$1 + 5 =$	$6 + 5 =$	$6 + \quad =11$
	$5 + 1 =$	$5 + 6 =$	$2 + \quad = 7$
	$2 + 5 =$	$7 + 5 =$	$8 + \quad =13$
	$5 + 2 =$	$5 + 7 =$	$4 + \quad = 9$
$5 + 5 =$	$3 + 5 =$	$8 + 5 =$	$1 + \quad = 6$
$6 + 5 =$	$5 + 3 =$	$5 + 8 =$	$5 + \quad =10$
$7 + 5 =$	$4 + 5 =$	$9 + 5 =$	$3 + \quad = 8$
$8 + 5 =$	$5 + 4 =$	$5 + 9 =$	$7 + \quad =12$
$9 + 5 =$	$5 + 5 =$	$0 + 5 =$	$9 + \quad =14$

1. Five and seven are how many?

2. Five and what make 12?

3. Seven and what make 12?

4. Nine and five are how many?

5. Eight and what make 13?

6. Five and what make 13?

7. Five and what make 14?

8. Five and what make 11?

9. Five and what make 10?

10. George has 6 nuts and John has 5. How many have both?

11. Mary has 7 flowers. How many more must she get to have 12?

12. Jane has learned five words. How many more must she learn to know 10?

1	5	2	5	3	5	4	5	5	6	5	7	5	8	5	9	5
5	1	5	2	5	3	5	4	5	5	6	5	7	5	8	5	9

Seat Exercise.

1 + 6 =	5 + 6 =	6 + = 9
6 + 1 =	6 + 5 =	6 + =12
2 + 6 =	7 + 6 =	6 + =11
6 + 2 =	6 + 7 =	6 + = 7
3 + 6 =	8 + 6 =	6 + =10
6 + 3 =	6 + 8 =	6 + = 8
4 + 6 =	9 + 6 =	6 + =15
6 + 4 =	6 + 9 =	6 + =13
6 + 6 =	0 + 6 =	6 + =14

6 + 6 =
7 + 6 =
8 + 6 =
9 + 6 =

1. Six and eight are how many?
2. Six and what make 14?
3. Eight and what make 14?
4. Six and seven are how many?
5. Six and what make 13?
6. Seven and what make 13?
7. Six and nine are how many?
8. Six and what make 15?
9. Nine and what make 15?

10. If John gives Mary eight flowers, how many must James give her so that she will have 14?

12. John gave Mary seven flowers and James gave her six. How many did she then have?

13. If John finds 9 eggs, how many must George find to make 15?

1	6	2	6	3	6	4	6	5	6	6	7	6	8	6	9	6
6	1	6	2	6	3	6	4	6	5	6	6	7	6	8	6	9

Seat Exercise.

	$1 + 7 =$	$5 + 7 =$	$7 + = 13$
	$7 + 1 =$	$7 + 5 =$	$7 + = 8$
	$2 + 7 =$	$6 + 7 =$	$7 + = 10$
	$7 + 2 =$	$7 + 6 =$	$7 + = 14$
	$3 + 7 =$	$8 + 7 =$	$7 + = 16$
	$7 + 3 =$	$7 + 8 =$	$7 + = 9$
$7 + 7 =$	$4 + 7 =$	$9 + 7 =$	$7 + = 11$
$8 + 7 =$	$7 + 4 =$	$7 + 9 =$	$7 + = 15$
$9 + 7 =$	$7 + 7 =$	$0 + 7 =$	$7 + = 12$

1. Eight and seven are how many?
2. Eight and what make 15?
3. Seven and what make 15?
4. Nine and seven are how many?
5. Nine and what make 16?
6. Seven and what make 16?
7. Seven and what make 14?
8. John has learned 7 words. How many more must he learn to know 14?
9. If Frank and Mary find 15 eggs, and Mary finds seven of them, how many does Frank find?

10. Peter and John have 16 apples, and Peter has 9 of them. How many has John?

Seat Exercise.

	1 + 8 =	5 + 8 =	8 + =14
	8 + 1 =	8 + 5 =	8 + =10
	2 + 8 =	6 + 8 =	8 + = 9
	8 + 2 =	8 + 6 =	8 + =12
	3 + 8 =	7 + 8 =	8 + =16
	8 + 3 =	8 + 7 =	8 + =11
	4 + 8 =	9 + 8 =	8 + =15
8 + 8 =	8 + 4 =	8 + 9 =	8 + =13
9 + 8 =	8 + 8 =	0 + 8 =	8 + =17

1. Eight and eight are how many?
2. Eight and what make 16?
3. Eight and nine are how many?
4. Eight and what make 17?
5. Nine and what make 17?

6. If there are eight girls and nine boys in the yard, how many are there in all ?

7. If there are 17 children in the yard and eight of them are girls, how many are boys ?

8. If there are 17 children in the yard and nine of them are boys, how many are girls ?

1	8	2	8	3	8	4	8	5	8	6	8	7	8	8	9	8
8	1	8	2	8	3	8	4	8	5	8	6	8	7	8	8	9

Seat Exercise.

$1 + 9 =$	$5 + 9 =$	$9 + \quad =11$
$9 + 1 =$	$9 + 5 =$	$9 + \quad =13$
$2 + 9 =$	$6 + 9 =$	$9 + \quad =15$
$9 + 2 =$	$9 + 6 =$	$9 + \quad =10$
$3 + 9 =$	$7 + 9 =$	$9 + \quad =12$
$9 + 3 =$	$9 + 7 =$	$9 + \quad =16$
$4 + 9 =$	$8 + 9 =$	$9 + \quad =14$
$9 + 4 =$	$9 + 8 =$	$9 + \quad =17$
$9 + 9 =$	$9 + 9 =$	$0 + 9 =$	$9 + \quad =18$

1. Nine and nine are how many ?

2. Nine and what make 18 ?

3. How many nines make 18 ?

4. There are 8 red apples and 9 green apples in a dish. How many apples are there in the dish ?

5. There are 17 apples in a dish. 9 of them are red and the others green. How many are green ?

1	9	2	9	3	9	4	9	5	9	6	9	7	9	8	9	9
9	1	9	2	9	3	9	4	9	5	9	6	9	7	9	8	9

New Class Exercise.

Purpose.—*To teach to recognize instantly the two parts which make each of the numbers from 2 to 10.*

Method.—Begin with 5, as this gives more variety than 2, 3, or 4. Show that $4+1$ or $1+4$, $3+2$ or $2+3$, make 5. Drill upon it until the pupils can recognize *at sight* the two component parts of 5. So teach the component parts of each of the other numbers from 2 to 10; thus, of 6 they are $5+1$ or $1+5$, $4+2$ or $2+4$, and $3+3$. Of 7 they are $6+1$ or $1+6$, $5+2$ or $2+5$, $4+3$ or $3+4$; etc.

Write these combinations promiscuously on the blackboard, and require the pupils to give the sum of any couplet instantly as you point to the couplet. Remember that to recognize the sum *at sight* is quite as important as to do it when the numbers are given orally, and that the pupil may do one readily and not the other.

Having a large number of figures on the blackboard or on a chart before the class, point to one figure, as 4, and then to another, as 3, and train the pupils till they can give the sum " as quick as thought."

An exercise like this will be very useful both for its own sake and as a preparation for subtraction: *Teacher,* " I will give one of the parts of 8 and the class may give the other." T., " 5 ; " C., " 3 ;" T., " 4 ;" C., " 4 ;" T., " 6 ;" C., " 2," etc.

Seat Exercise.

1. Write each two numbers which make 2.

2. Write each two numbers which make 3.

3. Write each two numbers which make 4. Each two which make 5. Each two which make 6 ; 7 ; 8 ; 9 ; 10.

LESSON III.

Purpose.—*To teach how to add any number expressed by two figures, to any one expressed by one figure.*

Method.—The first four exercises are learned by a simple recognition of the meaning of the words thirteen, fourteen, etc.

Class Exercise.

Write 10 in one place on the board and 7 in another. "What is this?" (Ten.) "What is this?" (Seven.) "Ten and seven are how many?" "What do we call ten and seven?" Question them until they recall the fact that ten and seven are (or are called) seventeen. Thus proceed with 10 and 8, 10 and 4, 10 and 9, etc. This is a review of the process of counting from ten to twenty, but is now to be seen in a slightly different light.

Write $10 + 1 =$, $10 + 4 =$, $10 + 6 =$, etc., and teach them to fill the blanks, and perform the following seat exercise.

Seat Exercise.

$10 + 1 =$	*$10 + 4 =$	$10 + \quad = 19$
$10 + 2 =$	$10 + 9 =$	$10 + \quad = 15$
$10 + 3 =$	$10 + 1 =$	$10 + \quad = 16$
$10 + 4 =$	$10 + 3 =$	$10 + \quad = 17$
$10 + 5 =$	$10 + 5 =$	$10 + \quad = 18$
$10 + 6 =$	$10 + 2 =$	$10 + \quad = 12$
$10 + 7 =$	$10 + 8 =$	$10 + \quad = 13$
$10 + 8 =$	$10 + 6 =$	$10 + \quad = 14$

---

* It is not necessary that these be taken in the reverse order, as $4 + 10$, $9 + 10$, etc., as the combinations do not occur in this order in ordinary addition.

1. There were ten eggs in the nest, and the hen laid three more. How many were there then?

2. John has ten books and Frank gives him five. How many books has he then?

3. Mary has ten flowers How many more must she pick to have 16?

4. There are 10 birds in the barn. How many more must come to make 18?

5. Ten and how many make 13?

6. Ten and how many make 17?

5	8	7	6	4	1	0	9	3	9
10	10	10	10	10	10	10	10	10	10

Recitation and Class Exercises.

The recitations and class exercises in this lesson will be similar to those in the last. Many other exercises like those in the Seat Exercises will be given orally. In general, such recitation and class exercise will consist, *First*, in an examination of slates, to see if all has been done correctly and neatly; *Second*, with books in hand, the pupils will read the seat exercises and tell the answers, one pupil reading the problem and another giving the answer; *Third*, other exercises will be dictated by the teacher; *Fourth*, concert exercises on the Addition Table, and exercises in adding numbers written on the blackboard.

Seat Exercise.

20+3=	30+5=	40+2=	50+8=
20+6=	30+7=	40+0=	50+1=
20+5=	30+9=	40+9=	50+0=
20+8=	30+0=	40+7=	50+7=
20+1=	30+1=	40+1=	50+6=
20+4=	30+3=	40+3=	50+3=
20+2=	30+6=	40+5=	50+4=
20+9=	30+2=	40+8=	50+2=
20+7=	30+8=	40+6=	50+6=
20+0=	30+4=	40+4=	50+5=

1. James has found 20 eggs, and Frank has found 7 more than James. How many has Frank found?

2. There are thirty blackbirds in a tree, and on the ground seven more than in the tree. How many birds are there on the ground?

3. There were 50 birds on the ground under a tree and 9 more came. How many birds were there then?

4. Ann's father was 40 years old, and Mary's father is 6 years older. How old is Mary's father?

5. In Frank's garden are 40 flowers; but in George's garden there are 7 more than in Frank's. How many flowers are there in George's garden?

	3	7	8	6	4	2	1	3	2	0
	50	40	20	30	20	50	40	30	50	20

Seat Exercise.

60 + 6 =	70 + 5 =	80 + 8 =	90 + 2 =
60 + 2 =	70 + 2 =	80 + 2 =	90 + 0 =
60 + 5 =	70 + 1 =	80 + 7 =	90 + 1 =
60 + 8 =	70 + 7 =	80 + 1 =	90 + 7 =
60 + 9 =	70 + 8 =	80 + 4 =	90 + 9 =
60 + 1 =	70 + 4 =	80 + 6 =	90 + 3 =
60 + 0 =	70 + 3 =	80 + 0 =	90 + 5 =
60 + 4 =	70 + 0 =	80 + 9 =	90 + 6 =
60 + 3 =	70 + 6 =	80 + 3 =	90 + 4 =
60 + 7 =	70 + 9 =	80 + 5 =	90 + 8 =

1. James has 80 nuts and finds 6 more. How many has he then?

2. If Mary has 70 flowers and Ann gives her 6 more, how many has she in all?

3. If George saw 90 birds and Frank saw 8 more than George did, how many did Frank see?

4. John has 10 cents and his father gives him 7 more. How many has he then?

5. James has 70 cents in a box and 5 cents in his hand. How many cents has he in all?

5	4	7	3	2	1	4	3	6	7	8
80	90	70	50	80	70	10	20	40	60	60

Second Class Exercise.

Purpose.—*To teach how to add any number represented by one digit to another represented by two digits, without counting.*

To teach how to find out how many 37 and 8 more make, write the numbers as in the margin. *First,* fix the attention upon the fact that 37 is 3 *tens* and 7. Use the picture or other objects. *Second,* 7 and 8 more make 15, which is 1 ten and 5. So we have 4 tens and 5, or 45. Use the picture or other objects. *There must be no counting.* The point is to show that *we have only to consider the sum of* TWO *digits, in any case.* In like manner illustrate the process with a variety of examples.

    8
   37
    45

This exercise need be continued only long enough to teach *how* the addition is effected. The two succeeding exercises exhibit the *expedients* by which *facility* is obtained.

Seat Exercise.

4	5	6	5	4	7	3	4	8	2
29	28	18	13	12	36	42	69	45	18

7	8	9	7	6	5	8	9	4	7
54	47	28	63	19	77	86	81	33	15

Third Class Exercise.

Purpose.—*To teach to recognize the sum of any digit added to any number represented by two digits, by remembering what digits when added give 0, 1, 2, 3, 4, 5, etc., in units place.*

Method.—Knowing that $9+1$, $8+2$, $6+4$, and $5+5$, each makes 10, the pupil is to be taught to recognize the sum in such cases as the following:

$$\frac{1}{19} \quad \frac{2}{18} \quad \frac{3}{17} \text{ etc. ;} \quad \frac{1}{29} \quad \frac{2}{28} \quad \frac{3}{27} \text{ etc. ;}$$

Of these there will be 81. The pupils may be required to write them all.

So, again, by knowing that $9+2$, $8+3$, $7+4$, and $6+5$, each makes 11 (or 1 in units place), the pupil is to be taught to recognize the sum in such cases as the following:

$$\frac{2}{19} \quad \frac{3}{18} \quad \frac{4}{17} \text{ etc. ;} \quad \frac{2}{29} \quad \frac{3}{28} \quad \frac{4}{27} \text{ etc. ;}$$

$$\frac{2}{39} \quad \frac{3}{38} \quad \frac{4}{37} \text{ etc. ;} \quad \frac{2}{49} \quad \frac{3}{48} \quad \frac{4}{47} \text{ etc. ;}$$

Of these combinations there are 72 in all. The pupils may be required to write them all as a seat exercise.

Another seat exercise may be obtained by requiring all the similar combinations which give 2 in the units place of the sum, as,

$$\frac{3}{19} \quad \frac{4}{18} \quad \frac{5}{17} \quad \frac{5}{16} \quad \frac{7}{15} \quad \frac{8}{14} \quad \frac{9}{13}$$

$$\frac{3}{29} \quad \frac{4}{28} \quad \frac{5}{27} \quad \frac{6}{26} \quad \frac{7}{25} \quad \frac{8}{24} \quad \frac{9}{23}$$

etc., etc., etc.

Of these there are 63.

Proceed in like manner to teach the combinations which give 3, 4, 5, 6, 7, 8, and 9, respectively, in the units place.

Fourth Class Exercise.

Purpose.—*This exercise is but a modified form of the preceding; and has the same end in view.*

Method.—Write on the blackboard a line like the following:

3	3	3	.3	3	3	3	3	3	3
4	14	24	34	44	54	64	74	84	94

Then call attention to the fact that as 4 and 3 make *seven*, 14 and 3 make *seven*-teen, 24 and 3 make twenty-*seven*, etc.; that is, that we have only to think what 4 and 3 make in any case. As a first exercise take only cases in which the tens are not changed, and give *several seat exercises* of this character.

Passing to the case in which the tens change, show the *reason* clearly as in the second exercise, but make it specially clear that the tens are only *one* more in any case, and that the important thing still is to recognize the sum of two figures. Finally, give seat and class exercises like the following till the idea is fixed.

2	2	2	2	2	2	2	2	2	2
3	13	23	33	43	53	63	73	83	93
5	5	5	5	5	5	5	5	5	5
4	14	24	34	44	54	64	74	84	94
5	5	5	5	5	5	5	5	5	5
6	16	26	36	46	56	66	76	86	96
8	8	8	8	8	8	8	8	8	8
4	14	24	34	44	54	64	74	84	94
7	7	7	7	7	7	7	7	7	7
6	16	26	36	46	56	66	76	86	96

There are 100 such combinations for each digit. Give thorough drill upon them, and conclude this lesson with miscellaneous exercises, continuing the lessons until the pupils can recognize the sum in any such case as readily as they can the sum of two digits.

New Class Exercise.

An excellent set of exercises can be supplied by the teacher thus :

Teach them to count *one hundred* by twos, as 2, 4, 6, 8, 10, etc., to 100. Then again, by beginning with *one*, count on by *twos*, as 1, 3, 5, 7, 9, 11, etc., to 99.

When this is learned, teach them to count by *threes*, first beginning with 3, then with 2, and then with 1. Thus they will count "3, 6, 9, 12, 15, etc., to 99;" or "2, 5, 8, 11, 14, etc.. to 98;" or "1, 4, 7, 10, 13, 16, etc., to 100."

Again, count by *fours*, first beginning with 4, then with 3, then with 2, then with 1.

Again, count by *fives*, by *sixes*, etc., to *nines*, in each case starting with each lower number.

This will readily be seen to be a most important exercise, and one that should be kept up for days, and perhaps for weeks, though it need not prevent the pupils from going on in the book, but may be made a frequent oral or slate exercise, as thought best by the teacher.

For Seat Exercises of this kind, tell them to write the numbers up to 100 by twos, as 2, 4, 6, 8, 10, etc., and so of all the other combinations here suggested.

Seat Exercise.

1. How is Mary counting when she says "3, 9, 15, 21, etc."? Count thus to 99.

2. How is one counting who says "2, 10, 18, 26, etc."? Count thus to 98.

3. How is John counting when he says "7, 16, 25, 34, etc."? Count thus to 97.

4. By what is James counting when he says "5, 9, 13, 17, etc."? Count thus to 97.

LESSON IV.

Purpose.—*To teach how to add any number of numbers expressed by one figure each, whose entire sum does not exceed one hundred.*

First Exercise.*

1. Here is an old barn, and in the yard are 4 hens with chickens. One hen has 5 chickens, another has 8, another has 6, and the other has 7. How many chickens are there in all? How many are $5 + 8 + 6 + 7$?

---

* Hereafter, suggestions upon class exercises and recitations will be put in foot-notes. *Every new process should be introduced by a familiar class exercise which will prepare the pupils to perform the seat exercise intelligently.*

The *class exercise* to be given before this *seat exercise* is assigned, will consist in showing with the counters, numeral frame, pictures such as those on pages 10, 11, 19, or other objects, how to add several numbers expressed by one figure each. Thus, suppose we wish to add 5, 4, 3, 7, and 6. Have one pupil

2. Mary has 4 flowers and Jane gives her 3 more. How many has she then? She then finds 2 more. How many has she then? When she brings them in, her mother gives her 1 more. How many has she in all? How many are 4 + 3 + 2 + 1?

3. How many are 4 and 8? How many are 12 and 5? How many are 17 and 6? Then how many are 4 + 8 + 5 + 6?

4. Here is a beautiful plant. On one stem are 5 flowers, on another 7, on another 6, and on another 8. How many flowers are there in all? How many are 5 and 7? How many are 12 and 6? How many are 18 and 8? Thus, how many are 5+7+6+8?

5. Eight apples in a dish, 4 on the table, 7 on the floor, and 9 in the chair, make how many apples?

6 + 4 + 8 + 5 + 3 = how many?

7 + 6 + 4 + 3 + 2 + 1 + 5 = how many?

			8	9	6	5
5	6	7	2	8	4	4
4	2	3	6	7	3	3
3	3	1	4	6	2	8
2	1	4	7	4	1	7
6	4	3	1	3	2	9

put 5 counters in a pile, another pupil 4 in another pile near the first, another pupil 3, another 7, and another 6. Then let them see clearly what the purpose is, viz.: to find how many there are in all *without counting them one by one.*

Second Exercise.

1. Frank bought a top for 8 cents, some nuts for 5 cents, a kite for 9 cents, a hoop for 7 cents, and a little book for 6 cents. How many cents did he pay for all ? $8 + 5 + 9 + 7 + 6 =$ how many ?

2. Mary bought a doll for 54 cents, a little book for 8 cents, and a hoop for 7 cents. How much did she pay for all ? $54 + 8 + 7 =$ how many ?

3. George bought a sled for 85 cents, a rope for 8 cents, an apple for 1 cent, and some nuts for 3 cents. How much did he pay for all ?

						4	6
		9	7	6	4	3	3
7	4	8	4	5	8	4	4
6	3	4	3	8	8	5	4
4	1	3	5	7	5	5	4
3	2	2	6	6	6	8	2
5	4	5	8	7	7	7	2
2	7	1	2	4	6	7	9
8	6	4	1	9	7	2	8

$3 + 4 + 5 + 8 + 7 + 5 + 4 + 9 =$ how many?
$5 + 8 + 2 + 1 + 1 + 3 + 3 + 4 =$ how many?

Put the 4 with the 5 and ask, "How many in this pile ?" (9.) Then the 3, asking as before ; then the 7, then the 6. Show also how to add the numbers when written in a column, and when written 5+4+3+7+6.

When the pupil hesitates in adding, as when he has 29 and the next figure is 7, ask, " 9 and 7 give what figures ? " thus teaching him to use the knowledge already gained. Pupils must be trained in this manner so that *they will not think of counting.*

These exercises are by no means sufficient to secure facility in adding. *Weeks of drill* are necessary. This may be kept up as a daily exercise while the pupil proceeds with other lessons.

LESSON V.

Definition Exercises.

Purpose.—*To teach the Meaning of the words Number, Add, Addition, Sum, and Amount, so that the pupil can understand them when used, and can use them.*

Method.—Ask questions involving the word, and if the child does not catch the meaning, put the question in a familiar form, and repeat the process till the purpose is accomplished.*

First Exercise.†

1. What number of boys do you see in the picture? What number of men? What number of trees?

---

* Formal definitions are out of place here, and all such questions as " What is number ? " " What is Addition ? " etc. To teach to *perceive* and to *conceive* constitute the purpose now ; to formulate thought is a later process, and to obtain ideas from formal statements and definitions a still later one.

† This must be preceded by an oral class exercise. 1. Print the word N u m b e r on the board, and teach it, if necessary. Then ask, " What *number*

2. How many ducks do you see in the picture? Ask the same question and use the word number. Ask the same question about the barrels and use the word number.

3. If you add the number of men in the picture to the number of boys, what number does it make? If you add the number of barrels to the number of ducks, what is the sum?

4. If you add the number of trees, barrels, and ducks, what is the sum? What other word can you use for sum?

5. Ask a question like the last about the trees, men, and boys.

6. What will be the sum if you add the number of boys and the number of ducks? Ask the same question about the trees and ducks. About the boys and trees.

7. If you add the number of your eyes, the number of your hands, and the number of your feet, what is the sum?

---

of hands have you, John?" "What *number* of fingers have you, Mary?" etc. If they do not answer readily, put the question thus: "*How many* hands have you?" Then put it as before. Then of objects out of sight. "What number of legs has a kitten?" etc. Again, put figures, as 5, 3, 10, 32, 87, etc., on the board, and ask, "What number is this?" "What number is this?" etc. Number means *how many*, is the child form of the thought.

2. "If I *add* the *numbers* 5 and 3, what *number* does it make?" So of other numbers. If they do not catch the meaning, ask "If I *put together* the numbers 5 and 3, how many does it make?" or, "What number does it make?"

3. "If I add 3 and 4, what is the *sum?*" or, "If I add 3 and 4, how many does it make?" etc. "If I add the *numbers* 7 and 8, what *number* is the *sum?*" etc. Use the word *amount* in the same manner. Sum or amount means *how many it makes*, will be the child thought.

4. "In all your lessons for some time you have been putting numbers together to find out how many they make. (Turn back and show them this.) This is called *Addition*. What have you been studying? What is putting numbers together to find how many they make called?"

Second Exercise.

1. What is the sum of 5, 8, 4, 7, and 6 ?
2. What is the sum of 27 and 8 ?
3. What is the amount of 15, 4, and 7 ?
4. What is the amount of 7, 9, 8, and 4 ?
5. What is the sum of 63 and 5 ?
6. What is the amount of 81 and 9 ?
7. What do you call finding the sum of several numbers ?
8. Add the numbers 8, 4, 7, 6, 5.
9. Add the numbers 10, 4, 8, 7, 9.
10. What number added to 5 makes 9 ?
11. What number added to 3 makes 7 ?
12. What number added to 7 makes 15 ?
13. What number added to 9 makes 13 ?
14. Find the sum of 8, 7, 6, 4, and 3.
15. Find the amount of 20, 6, 8, 4, and 2.
16. When you put several numbers together, what do you call the number which they make ?

SUBTRACTION.

Purpose.—*To teach how to recognize the remainder when any number less than 10 is taken from any number which is composed of that number and any number less than 10.*

Method.—*Consider what it takes with the given number to make the one from which it is to be taken.*

Illustrate with *Counters*, and with the *Numeral Frame*, that as 5 and 4 make 9, 4 from 9 leaves 5, and also that 5 from 9 leaves 4. Be sure that both facts are recognized.

First Exercise.

1. There are 3 pinks on a stock. How many will be left if you pick one? The one picked and the 2 left are how many? 1 and what make 3? 1 from 3 leaves how many? 3 less 1 is how many?

2. If you were to pick two of the pinks, how many would be left? The 2 picked and the 1 left are how many? 2 and what make 3? 2 from 3 leaves how many? 3 less 2 is how many?

3. If you have 7 apples in two piles, and there are 3 in one pile, how many are there in the other? 3 and what make 7? 3 from 7 leaves how many?

4. If you take 4 apples from a pile of 7 apples, how many will be left? Why? (Because 4 and 3 make 7.) 7 less 4 is how many? 7 less 3 is how many?

5. Two and 5 make how many? 2 from 7 leaves how many? Why? 5 from 7 leaves how many? Why?

6. A boy has 6 cents in one pocket and 3 cents in the other. How many has he in all? What do 6 and 3 make? If the boy loses the 3 cents out of one pocket, how many has he left? How many had he at first? How many did he lose? How many has he left? 3 from 9 leaves how many? Why?

1+ = 1	1 from 1 leaves how many?	1−1= *
1+ = 2	1 from 2 leaves how many?	2−1=
1+ = 3	1 from 3 leaves how many?	3−1=
1+ = 4	1 from 4 leaves how many?	4−1=
1+ = 5	1 from 5 leaves how many?	5−1=
1+ = 6	1 from 6 leaves how many?	6−1=
1+ = 7	1 from 7 leaves how many?	7−1=
1+ = 8	1 from 8 leaves how many?	8−1=
1+ = 9	1 from 9 leaves how many?	9−1=
1+ =10	1 from 10 leaves how many?	10−1=

1	2	3	4	5	6	7	8	9	10	1	2
1	1	1	1	1	1	1	1	1	1	0	0

7. Count backward from 10 to 0, by 1; thus, 10, 9, 8, etc. From 9 to 0. From 8 to 0.

8. Count backward from 7 to 0, by 1; thus, 7, 6, 5, etc. From 6 to 0. From 5 to 0. From 4 to 0. From 3 to 0.

This is an important *Drill Exercise*, and the teacher should be careful to have it thoroughly understood.

---

* Teach how to read this column. This and the corresponding columns in the subsequent exercises are to be thoroughly memorized.

Second Exercise.

1. There are 6 ducks in the pond. If 2 of them should come out, how many would remain? 2 and what are 6? 2 from 6 leaves how many? 6 less 2 is how many?

2. There are 5 eggs in the upper nest and 2 in the lower. How many more eggs are there in the upper than in the lower? 2 and how many more make 5? 2 from 5 leaves how many?

3. Mary has 7 cents and Frank has 2. How many more has Mary than Frank? 2 and how many more make 7? 2 from 7 leaves how many?

4. Henry is 2 years old. How many more years must he live to be 5 years old? 2 and how many make 5? 2 from 5 leaves how many?

5. Make 7 marks in a row on your slate. Then make two marks under them. How many marks have you in all? If you take away the 2 marks,

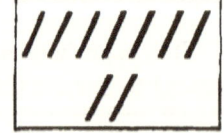

how many of the 9 marks will remain? 2 from 9 leaves how many? 2 and how many make 9? 9—2 is how many?

2+	= 2	2 from 2 leaves how many?	2—2=
2+	= 3	2 from 3 leaves how many?	3—2=
2+	= 4	2 from 4 leaves how many?	4—2=
2+	= 5	2 from 5 leaves how many?	5—2=
2+	= 6	2 from 6 leaves how many?	6—2=
2+	= 7	2 from 7 leaves how many?	7—2=
2+	= 8	2 from 8 leaves how many?	8—2=
2+	= 9	2 from 9 leaves how many?	9—2=
2+	=10	2 from 10 leaves how many?	10—2=
2+	=11	2 from 11 leaves how many?	11—2=

2	3	4	5	6	7	8	9	10	11	2	3
2	2	2	2	2	2	2	2	2	2	0	0

Drill Exercise.—*Conclude each of these lessons by drill exercises in counting backward;* thus, here have the pupils count backward from 11 to 0, by 2; then from 10; then from 9, etc. Conclude the next exercise by counting backward from 12 to 0, by 3, etc.

Third Exercise.

1. Little May is but 3 years old, and her brother Frank is 7 years old. How many years older is Frank than May? 3 and how many make 7? 3 from 7 leaves how many?

2. Make 9 dots on your slate, putting 3 in one group and 6 in another. 3 and how many make 9? 3 from 9 leaves how many?

3. A man had lost 3 fingers from one hand. How many had he left on that hand, counting the thumb?

4. There are 8 books on the table. How many will be left when Frank has taken the 3 small ones away? 3 and what make 8? 3 from 8 leaves how many?

3 + = 3	3 from 3 leaves how many?	3 — 3 =
3 + = 4	3 from 4 leaves how many?	4 — 3 =
3 + = 5	3 from 5 leaves how many?	5 — 3 =
3 + = 6	3 from 6 leaves how many?	6 — 3 =
3 + = 7	3 from 7 leaves how many?	7 — 3 =
3 + = 8	3 from 8 leaves how many?	8 — 3 =
3 + = 9	3 from 9 leaves how many?	9 — 3 =
3 + = 10	3 from 10 leaves how many?	10 — 3 =
3 + = 11	3 from 11 leaves how many?	11 — 3 =
3 + = 12	3 from 12 leaves how many?	12 — 3 =

3	4	5	6	7	8	9	10	11	12	3	5
3	3	3	3	3	3	3	3	3	3	0	0

Fourth Exercise.

1. John is now 10 years old, but he began going to school 4 years ago. How old was he when he began to go to school? 4 and what make 10? 4 from 10 leaves how many?

2. Little May is but 4 years old. How many years before she will be 11 years old ? 4 and what make 11 ? 4 from 11 leaves how many ?

3. Five tops and 4 tops are how many tops?

4. Nine tops are how many more than 5 tops?

5. Nine tops are how many more than 4 tops?

6. Mary has 13 cents and Carrie has 4. How many more has Mary than Carrie ?

4+ = 7	4 from 4 leaves how many ?	10−4=
4+ =10	4 from 5 leaves how many ?	8−4=
4+ = 4	4 from 6 leaves how many ?	4−4=
4+ =13	4 from 7 leaves how many ?	6−4=
4+ = 9	4 from 8 leaves how many ?	11−4=
4+ =11	4 from 9 leaves how many ?	9−4=
4+ = 5	4 from 10 leaves how many ?	5−4=
4+ = 8	4 from 11 leaves how many ?	7−4=
4+ = 6	4 from 12 leaves how many ?	13−4=
4+ =12	4 from 13 leaves how many ?	12−4=

11	6	5	4	12	10	13	8	7	9	4	5
4	4	4	4	4	4	4	4	4	4	0	0

Fifth Exercise.

1. How many birds are in the tree? How many are flying to the tree? How many more are there in the tree than flying to it? If as many should fly off from the tree as are flying to it, how many would be left in the tree? 8 less 5 is how many? 8 is how many more than 5?

2. If you have 5 apples, how many more must you get to have 12? 12 is how many more than 5? 5 from 12 leaves how many?

3. How many days in one week? After 5 days of a week are gone, how many are left? $7 - 5 =$ how many?

5+ = 8	5 from 5 leaves how many?	11−5=
5+ = 6	5 from 6 leaves how many?	6−5=
5+ =11	5 from 7 leaves how many?	8−5=
5+ = 9	5 from 8 leaves how many?	10−5=
5+ = 5	5 from 9 leaves how many?	13−5=
5+ = 7	5 from 10 leaves how many?	12−5=
5+ =12	5 from 11 leaves how many?	7−5=
5+ =14	5 from 12 leaves how many?	5−5=
5+ =10	5 from 13 leaves how many?	9−5=
5+ =13	5 from 14 leaves how many?	14−5=

12	8	7	10	5	9	6	11	13	14	5	6
5	5	5	5	5	5	5	5	5	5	0	0
—	—	—	—	—	—	—	—	—	—	—	—

Sixth Exercise.

1. If I buy a book for 6 cents, and hand the bookseller a dime (ten cents), how much change must he give me? 6 and what are 10? 6 from 10 leaves how many?

2. If I owe the postmaster 6 cents, and hand him 15 cents, how much change must he give me? 6 and what make 15? 6 from 15 leaves how many?

3. There were 15 peaches on the tree. How many are there now? Jane picked the others. How many did she pick?

4. There were 15 peaches on the tree, and Jane picked 6 of them. How many are left?

6+ =10	6 from 6 leaves how many?	6−6=
6+ = 7	6 from 7 leaves how many?	10−6=
6+ = 9	6 from 8 leaves how many?	15−6=
6+ =11	6 from 9 leaves how many?	13−6=
6+ = 6	6 from 10 leaves how many?	12−6=
6+ =12	6 from 11 leaves how many?	8−6=
6+ =14	6 from 12 leaves how many?	9−6=
6+ = 8	6 from 13 leaves how many?	7−6=
6+ =13	6 from 14 leaves how many?	11−6=
6+ =15	6 from 15 leaves how many?	14−6=

12	6	8	7	11	13	15	14	9	10	6	3
6	6	6	6	6	6	6	6	6	6	0	0

Seventh Exercise.

1. John had 11 cents, and bought a slate for 7 cents. How many cents did he have left? 7 and what make 11? 7 from 11 leaves how many?

2. Frank has 7 cents. How many more must he earn to have 13 cents?

3. Mary has 15 flowers and her brother Henry has 7. How many more has Mary than Henry?

4. From a flock of 12 chickens a fox caught 7. How many were left?

5. Frank had a ball worth 16 cents, and James had a top worth 7 cents. How much more is the ball worth than the top? If they trade, how many cents ought James to give to Frank besides giving him his top?

6. If a spool of thread is worth 7 cents, and I hand the merchant 10 cents for one, how much change must he give me?

7+ = 9	7 from 7 leaves how many?	7—7=
7+ =15	7 from 8 leaves how many?	10—7=
7+ =10	7 from 9 leaves how many?	16—7=
7+ = 8	7 from 10 leaves how many?	8—7=
7+ =11	7 from 11 leaves how many?	11—7=
7+ =13	7 from 12 leaves how many?	15—7=
7+ =12	7 from 13 leaves how many?	12—7=
7+ =14	7 from 14 leaves how many?	9—8=
7+ = 7	7 from 15 leaves how many?	13—7=
7+ =16	7 from 16 leaves how many?	14—7=

16	12	7	8	10	11	13	14	9	15	7	4
7	7	7	7	7	7	7	7	7	7	0	0

Eighth Exercise.

1. Mary had 13 cents, and bought a ribbon for 8 cents. How many cents had she left? 8 and what make 13 ?

2. John bought a lead-pencil for 8 cents, and handed the merchant 15 cents. How much change must he receive? 8 and what make 15 ? 8 from 15 leaves how many ?

3. When May had learned 17 words, her brother Frank had learned 8 less. How many had Frank learned ?

4. 6 and 8 are how many ? 6 from 14 leaves how many ? 8 from 14 leaves how many ? How would you illustrate this with the counters ?

8+ =12	8 from 8 leaves how many?	11−8=
8+ =10	8 from 9 leaves how many?	15−8=
8+ = 8	8 from 10 leaves how many?	17−8=
8+ = 9	8 from 11 leaves how many?	8−8−··
8+ =17	8 from 12 leaves how many?	10−8=
8+ =13	8 from 13 leaves how many?	12−8=
8+ =15	8 from 14 leaves how many?	14−8=
8+ =14	8 from 15 leaves how many?	9−8=
8+ =11	8 from 16 leaves how many?	13−8=
8+ =16	8 from 17 leaves how many?	16−8=

8	10	9	17	15	11	13	12	16	14	8	3
8	8	8	8	8	8	8	8	8	8	0	0

Ninth Exercise.

1. Make 17 marks on your slate in two rows, 9 in one row, and 8 in another row right under the first row. How many marks have you now in all? If you take the 9 away from the 17, how many will remain? If you take away the 8 from the 17, instead of the 9, how many will remain? 9 from 17 leaves how many? Why?

2. If you take 9 from 16, how many will remain? Why? How would you illustrate it with the counters?

3. James bought a book for 9 cents, which took all the money he had but 4 cents. How much did he have at first? 9 from 13 leave how many?

4. If John buys a ball for 9 cents, and hands the merchant 15 cents, how much change should he receive?

5. Mary is 18 years old, and her little sister Ann is but 9. How much older is Mary than Ann?

9+ =10	9 from 9 leaves how many?	12—9=
9+ =17	9 from 10 leaves how many?	9—9=
9+ =11	9 from 11 leaves how many?	11—9=
9+ =15	9 from 12 leaves how many?	18—9=
9+ =12	9 from 13 leaves how many?	16—9=
9+ = 9	9 from 14 leaves how many?	10—9=
9+ =13	9 from 15 leaves how many?	13—9=
9+ =18	9 from 16 leaves how many?	15—9=
9+ =16	9 from 17 leaves how many?	14—9=
9+ =14	9 from 18 leaves how many?	17—9=

17	9	10	15	18	12	14	16	11	13	9	5
9	9	9	9	9	9	9	9	9	9	0	0

Definition Exercise.*

1. There were 8 roses on the bush, and Mary has subtracted 3 of them. How many are there left? 3 subtracted from 8 leaves how many?

2. There are 7 eggs in the nest. If you subtract 3, how many will remain? If you subtract 4, how many will remain? If you subtract 3 from 7, what is the remainder? If you subtract 4 from 7, what is the remainder?

3. If you subtract 2 from 5, what is the remainder? If you subtract 3 from 5, what is the remainder?

4. If you subtract 5 from 9, what is the remainder? If you take 5 from 9, what number is left? (These questions mean the same thing.)

5. If you subtract 2 from 8, what is the remainder? If you take 2 from 8, what number is left?

6. If you subtract 3 from 9, what is the remainder? Ask this question without using the words subtract and remainder.

7. If you take 8 from 12, what number is left? Ask the same question and use the words subtract and remainder.

---

* For the general character and spirit of the oral exercise which precedes this, see foot-notes, pages 59, 60. *Subtract* and *remainder* are the words whose use is to be taught in this exercise.

8. Supply the proper words in the following:

If I —— 6 from 13, what is the —— ?

What is the —— when you —— 7 from 11 ?

When you —— 8 from 17, what is the —— ?

What is the —— when 9 is —— from 20 ?

Second Definition Exercise.

1. How many books are on the table? How many are on the chair? How many more books are there on the table than on the chair? What is the difference between 8 books and 5 books?

2. If an orange costs 7 cents and a lemon 5 cents, what is the difference between the price of an orange and the price of a lemon? What is the difference between 7 and 5?

3. What is the difference between 10 dollars and 6 dollars?

4. John worked 9 hours and Henry worked 5 hours. How many more hours did John work than Henry? What is the difference between 9 hours and 5 hours?

5. What is the —— between 11 and 6?

What is the —— when you take 6 from 11 ?

What is the —— between 8 and 3 ?

If you —— 7 from 15, what is the —— ?

How do you find the —— between 10 and 7 ?

Ans. I —— 7 from 10.

6. What number does 4 + 7 make? What number does 3 + 2 make? What is the difference between 11 and 5?

7. From the sum of 4 and 3 subtract 5.

8. From the ———— of 2, 5, and 6, subtract 8?

9. Add 3, 4, and 7, and from the amount subtract 9. What word could you use instead of amount in asking this question? What instead of subtract?

10. From 3 + 2 + 4 + 1 subtract 7. What number is 3 + 2 + 4 + 1?

11. What is the ———— between the ———— of 5, 2, and 1, and 3, 2, and 2? How much is 5+2+1? How much 3+2+2?

Drill Exercises.*

1. 4+3−2+6+1−8+5−2=how many?
2. 5−2+7−6−3+8−4+6=how many?
3. 1+2+3+4+5−9−2+7−6=how many?
4. 4+8+6−9−4−1+7−8+6=how many?
5. 13−6−4+7−5−8−3+7=how many?
6. 17−9−5+2+8−6+4−1=how many?
7. 7+8−9−2+6−7+2−3=how many?

---

* The teacher will need to explain fully, and illustrate by a number of examples, before the pupils are required to study these exercises. It is a most valuable exercise for class drill. Name numbers in this way, and let the pupils answer. Thus, *at first*, the teacher says, "4+2"—the pupils, in concert, "6;" the teacher, "plus 5"—the pupils, "11;" the teacher, "minus 4"—the pupils, "7," etc. After a little concert answering of this kind, *i. e.*, after each number, let all follow *silently*, as the teacher says, "2+3+4−6−1+8= how many?" Then all who know raise the hand. If their results do not agree, try it again and again. Such an exercise as this should be kept up for a week or two as the main exercise with the class, and should always be used with frequency as a general exercise for the school. For the more advanced, it may include multiplication and division.

8. $2+3+7-9+5-6+7-8=$ how many?
9. $6+7+5-9+3-8-2+7=$ how many?
10. $15-8-4-2+6-4-2-1=$ how many?
11. $16-7-6+4+2-6-3+9=$ how many?
12. $11+2-6-3+5+2-4+1=$ how many?
13. $10-3-2+7-8-3+6-4=$ how many?
14. $6+9+2-8-2+4-6+3=$ how many?
15. $14-6-5+3+2-3+2+4=$ how many?
16. $9+7-8-2+3-4-3+7=$ how many?
17. $8+6+2-7-5+3-1+3=$ how many?
18. $6+7+4-9-3+6-4+8=$ how many?
19. $3+8+2-7-6+4+5-9=$ how many?
20. $6-4+2+2+3+3-6-6=$ how many?

Practical Exercises.*

1. John had 5 cents and 6 cents. He then spent 8 cents, and afterward earned 4 cents. How many had he then?

2. Mary was very fond of flowers. She had 8 little plants, but 3 of them died. Then her cousin gave her 4 plants. How many had she at last

3. Henry had 15 cents, and spent 6 cents for an orange, 1 cent for a pencil, and 3 cents for some nuts. How many cents had he left?

---

* The teacher should illustrate these examples by using the counters for cents. Thus, for the first, put out five counters and then 6. Take away 8; then add 4. Question the pupils as to what the number is each time.

4. Frank earned 6 cents Monday, spent 4 cents Tuesday, earned 7 cents Wednesday, 4 cents Thursday, spent 5 cents Friday, earned 8 cents Saturday, and put 7 cents in the missionary-box on Sunday. How much of his week's earnings had he left?

5. There were 11 boys at play in the yard, when 5 of them went home, 2 went off to play with some other boys, and 4 new boys came. How many boys were there in the yard at last?

6. Frank found a hen's nest with 9 eggs in it. He took out 3, and two days after found that the hens had laid 5 more eggs in the nest. He then took 6 out of the nest. How many did he leave in the nest at last?

7. John has 17 cents. He lost 5, spent 4, earned 3, and gave away 6. How many had he then?

8. A man has agreed to work 9 hours. How many more hours has he to work after he has worked 5 hours? 9 less 5 is how many?

9. I bought an orange for 5 cents, and handed the grocer a piece of money worth 10 cents. How much change must he give me?

10. I gave a boy one dime, and he gave me a glass of chestnuts worth 8 cents. How many cents should he give me in change? 8 and how many make 10?

11. A man has 11 miles to ride. How many more has he to ride after he has ridden 6?

MULTIPLICATION.

Purpose.—*To teach how to find out the product of any number less than 11 multiplied by any number less than 11, and to fix the results in memory; i. e., to learn the multiplication table to 10 times 10.*

Method.—*Teach the pupil to find out the product by adding the number to itself the requisite number of times.*

First Exercise.*

1. If you pick a cherry and put it in your hand, and then pick another, how many cherries will you have? How many times have you picked 1 cherry? Two times 1 cherry are how many cherries?

2. $1 + 1 =$ how many? How many times 1 are $1 + 1$? Two times 1 are how many?

3. If you pick 1 cherry, then another, and then another, how many times will you have picked a cherry? How many cherries will you have? 3 times 1 cherry are how many cherries?

---

* Multiplication is to be taught as based on addition. *Counters* und the *Numeral Frame* will be of constant service. Explain the use of the sign ×, reading 3×4, "3 times 4," etc.

4. $1 + 1 + 1 =$ how many? How many times 1 are $1+1+1$? 3 times 1 are how many?

5. If your mother give you 1 cent each day for 4 days, how many times will she have given you 1 cent? How many cents will you have? 4 times 1 cent are how many cents?

6. $1+1+1+1 =$ how many? How many times 1 are $1+1+1+1$? 4 times 1 are how many?

7. If Jane breaks 1 needle each day, how many does she break in a week (6 days)? How many are 6 times 1?

8. The sign $\times$ means *times*, and we read 3×2, three times two.

9. Read $4 \times 1, 3 \times 1, 5 \times 2, 6 \times 4$.*

10. Read $3 \times 2, 4 \times 7, 5 \times 6, 8 \times 9, 7 \times 4$.

*1=	†$1 \times 1 =$
1+1=	$2 \times 1 =$
1+1+1=	$3 \times 1 =$
1+1+1+1=	$4 \times 1 =$
1+1+1+1+1=	·$5 \times 1 =$
1+1+1+1+1+1=	$6 \times 1 =$
1+1+1+1+1+1+1=	$7 \times 1 =$
1+1+1+1+1+1+1+1=	$8 \times 1 =$
1+1+1+1+1+1+1+1+1=	$9 \times 1 =$
1+1+1+1+1+1+1+1+1+1=	$10 \times 1 =$

1	1	1	1	1	1	1	1	1	1	0	0
1	2	3	4	5	6	7	8	9	10	1	4

* These paragraphs will need careful explanation before the pupil is required to study them. Show by the *Numeral Frame* what is meant by "3 times 1," etc.

† This column is to be copied by the pupil, the results written, and the whole thoroughly committed to memory.

Second Exercise.

1. There are 2 cherries in each bunch, and 2 bunches. How many times 2 cherries are there on the twig? How many cherries are there? 2 times 2 cherries are how many cherries?

2. $2+2=$ how many? How many times 2 is $2+2$? $2 \times 2 =$ how many?

3. John staid out of school 2 days to visit his uncle, 2 days because he was sick, and 2 days he played truant. How many times did he stay out 2 days? How many days did he stay out in all? 3 times 2 days are how many days?

4. *$2+2+2=$ how many? How many times 2 are $2+2+2$? 3×2 are how many?

5. Jane found 2 eggs on Monday, 2 on Tuesday, 2 on Wednesday, and 2 on Thursday. How many times did she find 2 eggs? How many did she find in all? 4 times 2 eggs are how many eggs?

6. *$2+2+2+2=$ how many? How many times 2 is $2+2+2+2$? $4 \times 2=$ how many?

7. 4 times 2 are how many? If 4 times 2 are 8, how many are 5 times 2? How many 2's must you take with 4 times 2, or 8, to make 5 times 2?

---

* The pupil is expected to perform the *addition*, thus keeping up a drill in adding.

8. 6 times 2 are 12. How many are 7 times 2? How many 2's must you put with 6 times 2, or 12, to make 7 times 2?

9. 8 times 2 are 16. How many are 9 times 2?*

$2+2=$	$2\times2=$
$2+2+2=$	$3\times2=$
$2+2+2+2=$	$4\times2=$
$2+2+2+2+2=$	$5\times2=$
$2+2+2+2+2+2=$	$6\times2=$
$2+2+2+2+2+2+2=$	$7\times2=$
$2+2+2+2+2+2+2+2=$	$8\times2=$
$2+2+2+2+2+2+2+2+2=$	$9\times2=$
$2+2+2+2+2+2+2+2+2+2=$	$10\times2=$

2	2	2	2	2	2	2	2	2	0	1	2
2	3	4	5	6	7	8	9	10	2	2	2

Third Exercise.

1. Frank spent 3 cents each day in the week except Sunday. How many times did he spend 3 cents? How many cents did he spend in

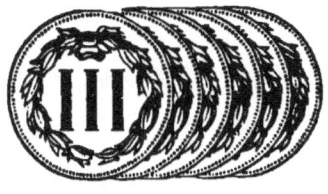

all? 6 times 3 cents are how many cents?

2. $3+3+3+3+3+3=$ how many? How many times 3 is $3+3+3+3+3+3$? 6×3 are how many?

---

* Special pains should be taken to have the pupil see the process as successive additions of the same number, so that if he knows what 7×6 is, he can tell what 8×6 is, etc.

3. George reads 3 pages each day of the week. How many times 3 pages does he read? How many are 7×3?

4. $3+3+3+3+3+3+3=$ how many? How many times 3 are $3+3+3+3+3+3+3$? 7 times 3 are how many?

5. James goes a fishing each day for 4 days and catches 0 fish each day. How many does he catch in all? How many are 4 times 0? 5 times 0?

6. 4 times 3 are 12. How many more 3's are 5 times 3 than 4 times 3? How many are 5 times 3?

7. 6 times 3 are 18. How many are 7 times 3?

8. $8 \times 3 = 24$. How many are 9×3? 10×3?

9. If you know how many 5 times 3 are, how can you tell from this how many 6 times 3 are?

$3+3+3=$	$3 \times 3 =$
$3+3+3+3=$	$4 \times 3 =$
$3+3+3+3+3=$	$5 \times 3 =$
$3+3+3+3+3+3=$	$6 \times 3 =$
$3+3+3+3+ +33+3=$	$7 \times 3 =$
$3+3+3+3+3+3+3+3=$	$8 \times 3 =$
$3+3+3+3+3+3+3+3+3=$	$9 \times 3 =$
$3+3+3+3+3+3+3+3+3+3=$	$10 \times 3 =$

$$\begin{array}{ccccccccccc} 3 & 3 & 3 & 3 & 3 & 3 & 3 & 3 & 0 & 3 & 3 \\ 3 & 4 & 6 & 8 & 7 & 4 & 5 & 8 & 5 & 7 & 10 \\ \hline \end{array}$$

$$\begin{array}{cccccccccc} 2 & 2 & 1 & 3 & 1 & 2 & 0 & 1 & 2 & 3 \\ 7 & 6 & 4 & 5 & 8 & 9 & 3 & 3 & 3 & 3 \\ \hline \end{array}$$

Fourth Exercise.

1. How many legs has 1 lamb? Five lambs have how many times as many legs as 1 lamb? How many legs have 5 lambs?

2. How many legs have 4 lambs? How many times as many legs as 1 lamb? How many are 4 times 4 ?

3. How many legs have 6 lambs? How many times as many legs as 1 lamb? Six times 4 are how many ?

4. Seven lambs have how many times as many legs as 1 lamb? 7 lambs have how many legs? $7 \times 4 =$ how many ?

5. James bought 8 oranges and gave 4 cents for each. How many times 4 cents did he give for all his oranges ? How many are 8 times 4 ?

6. $4+4+4+4+4+4+4+4+4 =$ how many ? How many times 4 are $4+4+4+4+4+4+4+4+4$? 9×4 $=$ how many?

7. 3 times 4 are 12. How many more 4's are 4 times 4 than 3 times 4 ? How many are 4×4 ?

8. $5 \times 4 = 20$. How many are 6×4 ? 7×4 ? 8×4 ? 9×4 ? 10×4 ? How many more do you take each time ?

4+4+4+4=	4×4=
4+4+4+4+4=	5×4=
4+4+4+4+4+4=	6×4=
4+4+4+4+4+4+4=	7×4=
4+4+4+4+4+4+4+4=	8×4=
4+4+4+4+4+4+4+4+4=	9×4=
4+4+4+4+4+4+4+4+4+4=	10×4=

4	4	4	4	4	4	4	3	3	3	2
5	7	4	6	8	10	9	3	7	4	7

1	3	2	3	1	3	0	1	2	3	4
8	6	8	5	5	8	4	4	4	1	1

Fifth Exercise.

1. How many points has one star ? How many have 5 stars? How many have 6 stars? 7 stars? 8 stars? 9 stars? 10 stars?

2. Six stars have how many times as many points as 1 star? Seven stars have how many times as many points as 1 star? Eight stars have how many times as many points as 1 star?

3. 5+5+5+5+5+5 are how many times 5? 6 times 5 are how many ?

4. $5+5+5+5+5+5+5+5$ are how many times 5? 8 times 5 are how many?

5. If John earns 5 cents each day, how many cents can he earn in 6 days? How many times as many can he earn in 6 days as in 1 day? Six times 5 are how many?

6. How many cents can John earn in 10 days, if he can earn 5 cents in 1 day? How many times as many cents can he earn in 10 days as in 1 day?

7. 3 times 5 are how many? How many more are 4 times 5? How many are 4 times 5? 7 times 5? 8 times 5? 9 times 5? 10 times 5? How many more 5's do you take each time?

$5+5+5+5+5=$	$5 \times 5=$
$5+5+5+5+5+5=$	$6 \times 5=$
$5+5+5+5+5+5+5=$	$7 \times 5=$
$5+5+5+5+5+5+5+5=$	$8 \times 5=$
$5+5+5+5+5+5+5+5+5=$	$9 \times 5=$
$5+5+5+5+5+5+5+5+5+5=$	$10 \times 5=$

5	5	5	5	5	5	0	1	2	3	4	5
5	6	7	8	9	10	5	5	5	5	5	5

4	3	2	3	4	5	1	0	4	4	4	4
7	7	8	8	8	8	8	8	6	9	10	4

Sixth Exercise.

1. How many petals* has one lily? How many times as many have 7 lilies? How many petals have 7 lilies? 7 times 6 are how many?

2. If 1 lily has 6 petals, how many petals have 8 lilies? How many have 9 lilies? How many times 6 petals have 10 lilies? How many petals have 10 lilies?

3. If James buys 6 oranges for 6 cents apiece, how many cents must he pay for all?

4. How many times as much must James pay for 9 oranges as for 1? How much must he pay for 9 oranges if 1 orange is worth 6 cents?

5. 6 + 6 + 6 + 6 + 6 + 6 are how many times 6? 6 times 6 are how many?

6. If each of 8 boys has 6 nuts, how many have they all? How many times 6 nuts have they? 8 times 6 nuts are how many nuts?

7. 2 times 6 are 12. How many more 6's must you take to make 3 times 6? How many are 3 times 6? How many more are 4 times 6? How many are 4 times 6?

---

* Teacher be careful to explain the meaning of this word, and teach the pupil how to pronounce it.

8. $5 \times 6 = 30$. How many are 6×6? 7×6?

9. $8 \times 6 = 48$. How many are 9×6? 10×6?

10. When you know how many 7 times 6 are, how do you find out how many 8 times 6 are?

$6+6+6+6+6+6=$	$6 \times 6 =$
$6+6+6+6+6+6+6=$	$7 \times 6 =$
$6+6+6+6+6+6+6+6=$	$8 \times 6 =$
$6+6+6+6+6+6+6+6+6=$	$9 \times 6 =$
$6+6+6+6+6+6+6+6+6+6=$	$10 \times 6 =$

6	6	6	6	6	0	1	2	3	4	5	6
9	10	7	6	8	6	6	6	6	6	6	6

Seventh Exercise.

1. There are 7 days in one week. How many times as many days are there in 4 weeks? How many days in 4 weeks? How many days in 2 weeks? How many in 6 weeks?

2. How many days are there in 8 weeks? Why?* How many days in 9 weeks? Why? How many days in 10 weeks? Why?

3. If Jane finds 7 eggs each day, how many will she find in 6 days? Why?

4. $7+7 =$ how many? $7+7$ are how many times 7? 2 times 7 are how many?

---

* The answer should be, " Because there are 8 times as many days in 8 weeks as there are in 1 week, and 8 times 7 are 56."

5. How many times 7 stars are there here? How many stars in a row from left to right? How many rows? How many times as many stars are there in 8 rows as there are in 1 row? 8 times 7 stars are how many stars?

6. How many stars are there in 7 of the rows from left to right? Why?

7. If 3 times 7 are 21, how many are 4 times 7 ?

8. If 5 times 7 are 35, how many are 6 times 7 ?

9. How many more are 8 times 7 than 7 times 7 ?

10. $4 \times 7 = 28$.   How many are 5×7? 6×7? 7×7?

$7+7+7+7+7+7+7=$	$7 \times 7=$
$7+7+7+7+7+7+7+7=$	$8 \times 7=$
$7+7+7+7+7+7+7+7+7=$	$9 \times 7=$
$7+7+7+7+7+7+7+7+7+7=$	$10 \times 7=$

7	7	7	7	0	1	2	3	4	5	6	7
8	9	7	10	7	7	7	7	7	7	7	7

2	3	4	5	6	7	8	7	5	4	6	3
3	4	5	6	7	4	3	7	5	4	6	3

Eighth Exercise.

1. How many o's are there in a row? How many e's? How many b's? How many letters in each row? How many rows? How many times 8 letters are there? How many letters in all? 9 times 8 letters are how many letters?

```
o o o o o o o o
e e e e e e e e
a a a a a a a a
i i i i i i i i
u u u u u u u u
c c c c c c c c
b b b b b b b b
f f f f f f f f
g g g g g g g g
```

2. If we write a row of 8 m's under the g's, how many rows of letters will there be? How many times 8 letters? 10 times 8 letters are how many letters?

3. If we cover up the row of g's, how many rows of letters will there be? How many times 8 letters? 8 × 8 = how many?

4. If there are 9 boys, and each boy has 8 cents, how many times 8 cents have they all? How many cents have they all? 9 times 8 are how many?

5. If there are 10 girls in the class, and each girl has 8 buttons on her dress, how many buttons are there on all their dresses? How many times 8 buttons are there? 10 times 8 are how many?

6. 2 × 8 = 16. How many are 3 × 8? 4 × 8? 5 × 8? 6 × 8? 7 × 8?

$$8+8+8+8+8+8+8+8 = \qquad 8 \times 8 =$$
$$8+8+8+8+8+8+8+8+8 = \qquad 9 \times 8 =$$
$$8+8+8+8+8+8+8+8+8+8 = \qquad 10 \times 8 =$$

8	8	8	0	1	2	3	4	5	6	7	8
10	8	9	8	8	8	8	8	8	8	8	8·

Ninth Exercise.

1. How many branches has this plant? How many leaves on each branch? How many times 9 leaves are there on the plant? 10 times 9 leaves are how many leaves?

2. If you were to break off the lowest branch, how many branches would be left? How many leaves? $9 \times 9 =$ how many?

3. If you were to break off two of the lower branches, how many branches would remain? How many leaves? $8 \times 9 =$ how many?

4. If you were to break off 3 of the branches, how many branches would remain? How many leaves? 7 times 9 are how many?

5. How many fingers has a boy on both hands, with his thumbs? How many have 10 boys? 10 tens make what?

6. How many fingers have 8 boys? 7 boys? 6 boys? 9 boys?

7. If a boy earns 10 cents each day, how many cents does he earn in 2 days? 3 days? 6 days? 8 days? 10 days? $10 \times 10 =$ how many?

8. $2 \times 9 = 18$. $3 \times 9 = ?$ $4 \times 9 = ?$ $5 \times 9 = 45$. $6 \times 9 = ?$ $7 \times 9 = ?$ $8 \times 9 = ?$

$$9+9+9+9+9+9+9+9+9= \quad | \quad 9\times9=$$
$$9+9+9+9+9+9+9+9+9+9= \quad | \quad 10\times9=$$
$$10+10+10+10+10+10+10+10+10+10=$$
$$10\times10=$$

9	9	0	1	2	3	4	5	6	7	8	9	10
9	10	9	9	9	9	9	9	9	9	9	9	9

10	0	1	2	3	4	5	6	7	8	9	10
10	10	10	10	10	10	10	10	10	10	10	10

Tenth Exercise.*

1. Here are two piles of money. In the upper there are 5 3-cent pieces. How many cents are in it? 5 times 3 are how many? In the lower are 3 5-cent pieces. How many cents in it? 3 times 5 are how many? In which pile is there the most money? 5 times 3 is the same as what?

2. James earned 3 cents each day for 4 days. How many cents had he? John earned 4 cents each day for 3 days. How many cents had he? Which had the more?

---

* The purpose of this exercise is to teach that *the factors may be exchanged without affecting the product ; i. e ,* that 4 times 3 is the same as 3 times 4, etc This truth should be amply illustrated by oral exercises akin to these here given, before the pupil is required to study these.

3. 4 times 3 are how many? 3 times 4 are how many? Which is the most, 3 times 4 or 4 times 3?

4. Mary has 10 5-cent pieces and Jane has 5 10-cent pieces. Which has the most? Why?

5. If you count the rows of stars from left to right, how many stars are there in a row? How many rows of 7 stars each? How many stars? If you count the rows of stars down the page, how many stars are there in a row? How many rows of 8 stars each? 8 times 7 are how many? 7 times 8 are how many? Which is the most, 8 times 7 or 7 times 8?

6. 5 times 3 are how many? Then how many are 3 times 5? Why?*

7. 6 times 4 are how many? Then 4 times 6 are how many? Why?

8. 7 times 6 are how many? Then 6 times 7 are how many? Why?

$*1 \times 2 =$, hence $2 \times 1 =$ | $1 \times 8 =$, hence $8 \times 1 =$
$2 \times 2 =$ | $2 \times 8 =$, hence $8 \times 2 =$
$1 \times 3 =$, hence $3 \times 1 =$ | $3 \times 8 =$, hence $8 \times 3 =$
$2 \times 3 =$, hence $3 \times 2 =$ | $4 \times 8 =$, hence $8 \times 4 =$
$3 \times 3 =$ | $5 \times 8 =$, hence $8 \times 5 =$
$1 \times 4 =$, hence $4 \times 1 =$ | $6 \times 8 =$, hence $8 \times 6 =$
$2 \times 4 =$, hence $4 \times 2 =$ | $7 \times 8 =$, hence $8 \times 7 =$
$3 \times 4 =$, hence $4 \times 3 =$ | $8 \times 8 =$
$4 \times 4 =$ | $1 \times 9 =$, hence $9 \times 1 =$
$1 \times 5 =$, hence $5 \times 1 =$ | $2 \times 9 =$, hence $9 \times 2 =$
$2 \times 5 =$, hence $5 \times 2 =$ | $3 \times 9 =$, hence $9 \times 3 =$
$3 \times 5 =$, hence $5 \times 3 =$ | $4 \times 9 =$, hence $9 \times 4 =$
$4 \times 5 =$, hence $5 \times 4 =$ | $5 \times 9 =$, hence $9 \times 5 =$
$5 \times 5 =$ | $6 \times 9 =$, hence $9 \times 6 =$
$1 \times 6 =$, hence $6 \times 1 =$ | $7 \times 9 =$, hence $9 \times 7 =$
$2 \times 6 =$, hence $6 \times 2 =$ | $8 \times 9 =$, hence $9 \times 8 =$
$3 \times 6 =$, hence $6 \times 3 =$ | $9 \times 9 =$
$4 \times 6 =$, hence $6 \times 4 =$ | $1 \times 10 =$, hence $10 \times 1 =$
$5 \times 6 =$, hence $6 \times 5 =$ | $2 \times 10 =$, hence $10 \times 2 =$
$6 \times 6 =$ | $3 \times 10 =$, hence $10 \times 3 =$
$1 \times 7 =$, hence $7 \times 1 =$ | $4 \times 10 =$, hence $10 \times 4 =$
$2 \times 7 =$, hence $7 \times 2 =$ | $5 \times 10 =$, hence $10 \times 5 =$
$3 \times 7 =$, hence $7 \times 3 =$ | $6 \times 10 =$, hence $10 \times 6 =$
$4 \times 7 =$, hence $7 \times 4 =$ | $7 \times 10 =$, henee $10 \times 7 =$
$5 \times 7 =$, hence $7 \times 5 =$ | $8 \times 10 =$, hence $10 \times 8 =$
$6 \times 7 =$, hence $7 \times 6 =$ | $9 \times 10 =$, hence $10 \times 9 =$
$7 \times 7 =$ | $10 \times 10 =$

---

* This table affords exercises like the preceding, and should be used to familiarize the idea that the order of the factors is indifferent, and also as an exercise to aid in fixing the products in mind. Pupils should copy it and fill it out; they should recite it individually and in concert.

Eleventh Exercise.

1. *6+6+6+6 are how many times 6 ?   How many are 4 times 6 ?

2. 3+3+3+3+3+3+3+3 are how many?   How many times 3 ?   8 times 3 are how many ?

1+ 1= ,hence 2 × 1=	6+ 6= ,hence 2 × 6=
2+ 2= ,hence 2 × 2=	7+ 7= ,hence 2 × 7=
3+ 3= ,hence 2 × 3=	8+ 8= ,hence 2 × 8=
4+ 4= ,hence 2 × 4=	9+ 9= ,hence 2 × 9=
5+ 5= ,hence 2 × 5=	10+10= ,hence 2 ×10=

$$1 + 1 + 1 = \quad , \text{ hence } 3 \times 1 =$$
$$2 + 2 + 2 = \quad , \text{ hence } 3 \times 2 =$$
$$3 + 3 + 3 = \quad , \text{ hence } 3 \times 3 =$$
$$4 + 4 + 4 = \quad , \text{ hence } 3 \times 4 =$$
$$5 + 5 + 5 = \quad , \text{ hence } 3 \times 5 =$$
$$6 + 6 + 6 = \quad , \text{ hence } 3 \times 6 =$$
$$7 + 7 + 7 = \quad , \text{ hence } 3 \times 7 =$$
$$8 + 8 + 8 = \quad , \text{ hence } 3 \times 8 =$$
$$9 + 9 + 9 = \quad , \text{ hence } 3 \times 9 =$$
$$10 + 10 + 10 = \quad , \text{ hence } 3 \times 10 =$$

3. 1+1+1+1=   , hence 4×1=
2+2+2+2=   , hence 4×2=
3+3+3+3=   , hence 4×3=
4+4+4+4=   , hence 4×4=

Let the pupil copy this on his slate and fill it out to 4 times 10, and write in all the results. So also of the following:

---

* The pupil is expected to *add* the 6's, and thus find out the answer. It is designed that exercise in addition as well as instruction in multiplication be secured.

4. $1+1+1+1+1=$, hence $5 \times 1=$
$2+2+2+2+2=$, hence $5 \times 2=$, etc.

5. $1+1+1+1+1+1=$, hence $6 \times 1=$
$2+2+2+2+2+2=$, hence $6 \times 2=$, etc.

6. $1+1+1+1+1+1+1=$, hence $7 \times 1=$
$2+2+2+2+2+2+2=$, hence $7 \times 2=$, etc.

7. $1+1+1+1+1+1+1+1=$, hence $8 \times 1=$
$2+2+2+2+2+2+2+2=$, hence $8 \times 2=$, etc.

8. $1+1+1+1+1+1+1+1+1=$, hence $9 \times 1=$
$2+2+2+2+2+2+2+2+2=$, hence $9 \times 2=$, etc.

9.
$1+1+1+1+1+1+1+1+1+1=$, hence $10 \times 1=$
$2+2+2+2+2+2+2+2+2+2=$, hence $10 \times 2=$, etc.

Twelfth Exercise.

1. Repeat the 2's of the multiplication table 5 times, thus: *

1 time 2 is ——.
2 times 2 are ——.
3 times 2 are ——, etc.

2. Repeat the 2's 5 times in this way:

2 times 1 are ——.
2 times 2 are ——.
2 times 3 are ——, etc.

---

* Teacher show the child how to keep his tally by marks as he says the exercise over, so as to know when he has been over it five times.

3. Repeat the 3's 5 times thus:

1 time 3 is ——.

2 times 3 are ——.

3 times 3 are ——, etc.

4. Repeat the 3's 5 times thus:

3 times 1 are ——.

3 times 2 are ——.

3 times 3 are ——, etc.

5. Answer the following 5 times:

2×3?　4×2?　2×7?　7×2?　3×1?　6×3?　5×3?

3×5?　4×3?　3×4?　2×8?　8×2?　7×3?　2×7?

1×2?　1×3?　5×2?　3×5?　9×3?　9×2?　3×9?

6. If 1 orange costs 4 cents, how many cents will 3 oranges cost?

If 1 orange cost 4 cents, 3 oranges will cost 3 times 4 cents, or 12 cents.

7. If 1 pencil costs 5 cents, how many cents will 3 pencils cost?

8. If a boy learns 2 lessons each day, how many lessons does he learn in 6 days? In 9 days?

9. There are 7 days in one week. How many days are there in 2 weeks? In 3 weeks?

Thirteenth Exercise.

1. Repeat the 4's 5 times in each of the two ways, thus:

1 time 4 is ——.	4 times 1 are ——.
2 times 4 are ——.	4 times 2 are ——.
3 times 4 are ——.	4 times 3 are ——.
etc., etc.	etc., etc.

2. Repeat the 5's 5 times in each of the two ways, thus:

1 time 5 is ——.	5 times 1 are ——.
2 times 5 are ——.	5 times 2 are ——.
3 times 5 are ——.	5 times 3 are ——.
etc., etc.	etc., etc.

3. Repeat the 6's 5 times in each of the two ways, thus:

1 time 6 is ——.	6 times 1 are ——.
2 times 6 are ——.	6 times 2 are ——.
3 times 6 are ——.	6 times 3 are ——.
etc., etc.	etc., etc.

4. Copy the following on your slates, multiply, and write the results underneath :

6	4	7	8	9	6	7	8	5	4	6	9
4	7	5	6	6	7	6	5	8	9	9	6

7	5	9	6	6	4	5	10	2	6	6	5
6	7	4	10	8	10	10	5	4	2	6	5

5. James worked 7 hours for 5 cents an hour. How much did he earn ?

6. John worked 6 hours for 4 cents an hour, and Henry worked 4 hours for 6 cents an hour. Which earned the most ?

7. Jane bought 7 oranges for 6 cents each. How much did they cost?

8. Mary bought 4 spools of thread for 5 cents a spool, and gave the clerk 25 cents. How much change should he give her? How much did her thread cost? 25 is how much more than 20 ?

Fourteenth Exercise.

1. Repeat the 7's 5 times in each of the two ways, thus:

1 time 7 is ——.	7 times 1 are ——.
2 times 7 are ——.	7 times 2 are ——.
etc., etc.	etc., etc.

2. Repeat the 8's 5 times in each of the two ways, thus:

1 time 8 is ——.	8 times 1 are ——.
2 times 8 are ——.	8 times 2 are ——.
etc., etc.	etc., etc.

3. Answer the following 5 times:

8×7? 8×9? 6×8? 7×6? 8×6? 9×8? 9×7?
7×9? 6×9? 9×6? 7×7? 8×8? 9×9? 7×5?
7×4? 8×3? 3×8? 6×3? 4×6? 3×7? 3×9?

4. If 7 white hens have 8 chickens each, and 8 black hens have 7 chickens each, which have the most chickens, the white hens or the black hens? Why?

5. John earns 6 cents an hour and works 7 hours, and Henry earns 7 cents an hour and works 6 hours. Which earns the most money?

6. Which costs the most, 8 oranges at 9 cents each, or 9 oranges at 8 cents each? Why?

7. In the first column there are 4 words, with 7 letters in each word, and in the second are 7 words, with 4 letters in each word. In which column are there the most letters? Why?

a n s w e r s	l a t e
s c h o l a r	g o e s
f o l l o w s	s a i l
c h i c k e n	c o a t
	f i n d
	s n o w
	r a i n

Fifteenth Exercise.

1. Repeat the 9's 5 times in each of the two ways, thus:

1 time 9 is ——.	9 times 1 are ——.
2 times 9 are ——.	9 times 2 are ——.
etc., etc.	etc., etc.

2. Repeat the 10's 5 times in each of the two ways, thus:

1 time 10 is ——.	10 times 1 are ——.
2 times 10 are ——.	10 times 2 are ——.
etc., etc.	etc., etc.

3. Answer the following 5 times. Write them on your slates in the same way as those on page 97.

9×3? 4×9? 10×7? 8×9? 6×9? 7×9? 9×6?
10×3? 3×10? 9×8? 9×6? 9×7? 3×9? 1×9?

4. Answer the following:

$3 \times 5 =$	$3 \times 7 =$	$2 \times 2 =$
$7 \times 8 =$	$3 \times 2 =$	$3 \times 3 =$
$6 \times 7 =$	$5 \times 8 =$	$4 \times 4 =$
$9 \times 8 =$	$9 \times 4 =$	$5 \times 5 =$
$4 \times 7 =$	$9 \times 3 =$	$6 \times 6 =$
$7 \times 5 =$	$4 \times 9 =$	$7 \times 7 =$
$6 \times 9 =$	$3 \times 9 =$	$8 \times 8 =$
$10 \times 8 =$	$2 \times 8 =$	$9 \times 9 =$
$5 \times 7 =$	$3 \times 8 =$	$10 \times 10 =$
$7 \times 6 =$	$10 \times 4 =$	$1 \times 1 =$
$9 \times 6 =$	$3 \times 10 =$	$5 \times 0 =$
$8 \times 7 =$	$10 \times 10 =$	$0 \times 3 =$

Sixteenth Exercise.

1. How many times must you make
3 stars to have 12 stars?   How many
times 3 is 12 ?   4 times 3 are how many ?

```
 *    *    *
* *  * *  * *
```

2. How many times must you make
4 marks to have 20 marks ?   How many
times 4 is 20 ?   5 times 4 are how many ?

```
//// //// //// ////
```

3. Three times what number makes 12 ?   Three times
what number makes 18 ?

4. 4 times what number makes 20 ?
   5 times what number makes 30 ?
   7 times what number makes 21 ?
   8 times what number makes 56 ?

5. Copy the following table on your slates and fill it out :*

$2 \times$	$= 2$	$3 \times$	$=21$	$4 \times$	$=20$
$2 \times$	$= 6$	$3 \times$	$=15$	$4 \times$	$=16$
$2 \times$	$= 8$	$3 \times$	$=18$	$4 \times$	$= 4$
$2 \times$	$= 4$	$3 \times$	$=12$	$4 \times$	$=12$
$2 \times$	$=10$	$3 \times$	$= 3$	$4 \times$	$=24$
$2 \times$	$=14$	$3 \times$	$= 9$	$4 \times$	$=40$
$2 \times$	$=20$	$3 \times$	$=27$	$4 \times$	$= 8$
$2 \times$	$=18$	$3 \times$	$= 6$	$4 \times$	$=36$
$2 \times$	$=16$	$3 \times$	$=30$	$4 \times$	$=28$
$2 \times$	$=12$	$3 \times$	$=24$	$4 \times$	$=32$

6. How many times 5 does it take
to make 15 ?   How many 5's are
there in 15 ?

```
* *  * *  * *
 *    *    *
* *  * *  * *
```

7. How many times 6 does it take to make 24 ?   How
many 6's in 24 ?

---

* The teacher may need to explain how it is to be done.

Seventeenth Exercise.

1. How many cents do 6 5 cent pieces make? How many 5-cent pieces does it take to make 30 cents? 6 times 5 are how many? 6 times what number makes 30?

2. How many cherries are there in the picture? How many bunches? How many in each bunch? 7 times what number makes 42?

3. Copy the following table on your slates and fill it out:

5 ×	=15	6 ×	=18	7 ×	=28
5 ×	=10	6 ×	=12	7 ×	=14
5 ×	=20	6 ×	=42	7 ×	=35
5 ×	=35	6 ×	=48	7 ×	=63
5 ×	=50	6 ×	=24	7 ×	=70
5 ×	=45	6 ×	=30	7 ×	=42
5 ×	= 5	6 ×	= 6	7 ×	=21
5 ×	=25	6 ×	=54	7 ×	= 7
5 ×	=30	6 ×	=36	7 ×	=56
5 ×	=40	6 ×	=60	7 ×	=49

4. How many times must John bring in 4 eggs at a time, in order to bring in 24 eggs? How many times 4 does it take to make 24?

5. If Henry earns 5 cents an hour, how many cents will he earn in 7 hours?

6. If Henry earns 5 cents an hour, how many hours will it take him to earn 35 cents? Why? *Answer.* Because 7 times 5 are 35.

Eighteenth Exercise.

1. Here are two squares of black glass, with 9 flakes of snow on each. How many squares would it take to have 72 flakes of snow? 8 times 9 are how many?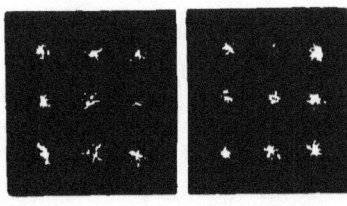

2. A class of boys had 90 fingers, including their thumbs. How many boys were there in the class. How many 10's does it take to make 90 ? 9 times 10 are how many?

3. Copy the following table on your slates and fill it out:

8 × = 16	9 × = 36	10 × = 40
8 × = 40	9 × = 27	10 × = 70
8 × = 24	9 × = 18	10 × = 90
8 × = 56	9 × = 45	10 × = 30
8 × = 64	9 × = 63	10 × = 10
8 × = 8	9 × = 81	10 × = 20
8 × = 72	9 × = 9	10 × = 100
8 × = 32	9 × = 54	10 × = 50
8 × = 80	9 × = 90	10 × = 80
8 × = 48	9 × = 72	10 × = 60

4. If John catches 7 fishes each day, how long will it take him to catch 42 fishes? 6 times 7 are how many?

5. If Moses lays by 10 cents a week, how many weeks will it take him to lay by one dollar, or 100 cents? How many tens does it take to make 100?

Definition Exercise.

1. If one currant bush produces two quarts of currants, how many quarts will 3 currant bushes produce? What is the product of 3 times 2? What is the product of 3 times 7? What is the product of 4 times 5?

2. What number does 6 times 4 make? What is the product of 6 times 4? What is the product of 6 times 7? What is the product of 7 times 6?

3. What is the product of 4 times 8? Ask this question without using the word product.

4. What is the product of 3 and 4? Ask this question without using the word product.

5. How many are 8 times 6? Ask this question and use the word product.

6. What is the product of 7 and 3?
What is the product of 6 and 9?
What is the product of 8 and 7?
What is the product of 3 and 9?
What is the product of 3 and 3?
What is the product of 8 and 8?

7. Supply the proper word in the following:
What is the ——— of 6 and 4?
The ——— of 7 and 5 is what?
9 times 5 gives what ———?
What is the ——— of 2 and 9?

———

Second Definition Exercise.

1. Here are some curious onions. If you plant a little one like one of those in the first row, it will grow and multiply into 4 or more like those in the second row. If you planted the 3 in the first row, and each one multiplied so as to make 4, how many would you have? 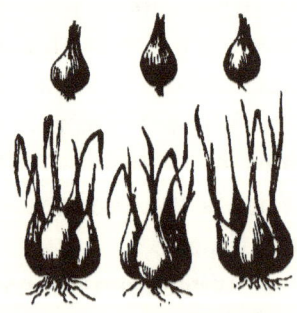 How many times as many as you planted? 3 multiplied by 4 produces how many? What is the product of 3 multiplied by 4?

2. If I plant 6 of these curious onions, and each one multiplies into 5, how many times as many shall I have as I planted? 6 multiplied by 5 produces how many?

3. If I multiply 5 by 6, what is the product? When I multiply 6 by 5, what is the product? When I multiply 5 by 6, how many times do I take 5? 6 times 5 are how many?

4. If I multiply 4 by 8, what is the product? Ask this question without using either of the words multiply or product.

5. If I —— 7 by 3, what is the ——?
If I —— 6 by 9, what is the ——?
The —— of 5 —— by 8 is what?
What is the —— of 7 —— by 6?

6. When you multiply 8 by 9, how many times do you take 8? What is the ——?

7. What is the —— of 8 and 7?
6 —— by 4 gives what ——?
What —— does 4 —— by 10 give?

8. Finding what the product of two numbers is, is called MULTIPLICATION.

9. We have now studied COUNTING, ADDITION, SUBTRACTION, and MULTIPLICATION. When I find out that 6 taken from 11 leaves 5, what is it? When I find out that 5 times 6 is 30, what is it? When I name all the numbers in order from one to twenty—thus, one, two, three, four, etc.—what is it? When I find out that 7 and 8 are 15, what is it?

Drill Exercise.*

1. Add 4 and 3 and 6; from this sum subtract 8; multiply this remainder by 2. What is the result?

2. Add 2 and 6 and 3 and 7; from this sum subtract 9; multiply the remainder by 3. What is the product?

3. Add 5 and 7; from this sum subtract 8; to this remainder add 5; from this sum subtract 7; multiply this remainder by 4; multiply this product by 3; to this product add 6. What is the result?

4. From 8 subtract 3; from this remainder subtract 2; to this remainder add 5 and 7; from this sum subtract 8; to this remainder add 2; multiply this sum by 4. What is the result?

5. Begin with 5, add 2, add 6, subtract 9, multiply by 3, add 5, subtract 8, subtract 6, multiply by 7. What is the result?

6. Begin with 11, subtract 6, subtract 2, multiply by 8, add 5. What is the result?

7. Begin with 6, multiply by 2, subtract 5, add 3, subtract 4, add 1, subtract 6, add 8, multiply by 7. What is the result?

8. Begin with 5, add 7, subtract 3, subtract 4, multiply by 3, add 8. What is the result?

9. Begin with 4, add 9, subtract 7, multiply by 2, subtract 1, subtract 3, multiply by 9. What is the result?

10. Begin with 6, add 8, subtract 7, add 2, multiply by 8. What is the result?

---

* Drill exercises of this character must be continually kept up as oral exercises. See foot-note on page 76.

Practical Exercises.

1. John bought 2 oranges for 4 cents each, and gave the clerk 10 cents. How much change did he receive?

2. Mary bought 3 spools of thread for 6 cents each, and one yard of calico for 9 cents. How much did she pay for all?

3. Henry worked 3 hours for 5 cents an hour, and the man for whom he worked gave him a ball worth 8 cents, and the remainder in money? How much money did Henry get?

4. How many days are there in one week? How many in 6 weeks? How many days in 4 weeks?

5. John worked 1 week (6 days) and 4 days more? How many days did he work in all? If he earned 4 shillings a day, how many shillings did he earn? How many more days would he have had to work to make 2 weeks? How much more would he have earned if he had worked 2 weeks?

6. Sarah sews 3 hours each day. How many hours does she sew in a week (6 days)?

7. How many days are there in 8 weeks? How many Sundays in 8 weeks? How many work-days in 8 weeks?

DIVISION.

Purpose.—*To develop the idea of Division in its Two Forms,* and the nature of Division as the converse of Multiplication, and to deduce the quotient of any number less than 100, divided by any number less than 10, from the relation of Division to Multiplication.*

First Exercise.†

1. How many little ducks are there in the picture? Are they all together? In how many groups are they? How many in each group? 3 groups,  with 4 in each group, make how many? How many 4's in 12?

2. If you have 12 little ducks and put them in 3 groups, with the same number in each group, how many will

---

* There are two essentially different logical processes called *division :* 1. Determining how many times one number is contained in another; and 2. Separating a number into any required number of equal parts, for the purpose of finding how many there are in one of these parts. The former is the more comprehensive view, although the latter gives name to the process. The foundation for Division has been so well laid in Multiplication that but little more will be needed here than to familiarize the *two forms* of conception, and give practice to fix the division table in mind.

† The teacher should carefully observe the character and purpose of the introductory examples in each of these exercises in Division, and give ample examples of *the same kind*, as class exercises, illustrating with the counters and other objects. But he sure and stick to the point of the particular exercise. Such questions as "How many 2's make 4?" "How many 3's make 15?" etc., are of great service in leading the pupil to comprehend the nature of division, and its relation to multiplication.

there be in each group? 12 divided into 3 equal parts makes how many in each part?

' 3. How many a's are there in the next line?

a   a   a   a   a   a

If you divide these 6 a's into groups with 3 in a group, how many groups will there be? | *a a a* | *a a a* |. How many 3's in 6? If you divide 6 a's into 2 equal groups, how many will there be in each group? 6 divided by 2 are how many?

4. 3 times what number makes 12? How many times does 4 go in 12? 4 times what number makes 12? How many times does 3 go in 12? 12 divided by 4 are how many? Why?* 12 divided by 3 are how many? Why?

5. 2 times what number make 6? How many are 6 divided by 3? Why? 3 times what number are 6? 6 divided by 2 are how many?

(NOTE.—Teacher, explain that ÷ means " divided by.")

6. Copy, fill out, and learn the following:

2÷2=	3÷3=	8÷2=	4÷2=
4÷2=	6÷3=	9÷3=	30÷3=
6÷2=	9÷3=	10÷2=	14÷2=
8÷2=	12÷3=	6÷3=	16÷2=
10÷2=	15÷3=	6÷2=	21÷3=
12÷2=	18÷3=	12÷3=	2÷2=
14÷2=	21÷3=	12÷2=	3÷3=
16÷2=	24÷3=	15÷3=	27÷3=
18÷2=	27÷3=	20÷2=	18÷2=
20÷2=	30÷3=	18÷3=	24÷3=

* The point of this question is that the pupil may learn to deduce Division from Multiplication. *Ans.* Because 3 times 4 are 12.

Second Exercise.

1. How many fingers, including thumbs, have two boys? How many 5's in 20? If you divide 20 into 5 equal parts, how many will there be in each part?

2. $20 \div 5 =$ how many?

3. $20 \div 4 =$ how many?

4. How many legs have 6 cats? How many 4's in 24? $24 \div 4 =$ how many? Why?

5. John has 28 cents; how many lemons can he buy at 4 cents each? How many 4's in 28? $28 \div 4 =$ how many? Why?

6. Mary has 15 pansies, and she wishes to make 5 bouquets and put the same number of pansies in each. How many can she put in each bouquet? If you divide 15 things into 5 equal groups, how many will there be in each group? $15 \div 5 =$ how many?

7. Copy, fill out, and learn the following:

$4 \div 4 =$	$5 \div 5 =$	$12 \div 4 =$	$4 \div 4 =$
$8 \div 4 =$	$10 \div 5 =$	$15 \div 5 =$	$5 \div 5 =$
$12 \div 4 =$	$15 \div 5 =$	$16 \div 4 =$	$10 \div 5 =$
$16 \div 4 =$	$20 \div 5 =$	$28 \div 4 =$	$35 \div 5 =$
$20 \div 4 =$	$25 \div 5 =$	$25 \div 5 =$	$32 \div 4 =$
$24 \div 4 =$	$30 \div 5 =$	$30 \div 5 =$	$40 \div 5 =$
$28 \div 4 =$	$35 \div 5 =$	$36 \div 4 =$	$40 \div 4 =$
$32 \div 4 =$	$40 \div 5 =$	$20 \div 4 =$	$45 \div 5 =$
$36 \div 4 =$	$45 \div 5 =$	$20 \div 5 =$	$24 \div 4 =$
$40 \div 4 =$	$50 \div 5 =$	$8 \div 4 =$	$50 \div 5 =$

Third Exercise.*

1. This large basket contains 42 eggs. How many times can the little girl fill her small basket from it, if her small basket holds 6 eggs? How many times can she fill her small basket if it holds 7 eggs? How many 6's in 42? How many times 6 make 42? 42 ÷ 6  are how many? Why? How many 7's in 42? How many times 7 make 42? 42÷7 make how many? Why?

2. Make 30 **O**'s on your slate, thus :

O O O O O O O O O O O O O O O O O O O O O O O O O O O O O O

Then rub out 6 **O**'s. Then rub out 6 more. Then another 6. How many times can you rub out 6 **O**'s? How many 6's in 30? How many times 6 make 30? 30÷6 are how many? Why?

3. If John has 35 cents and spends 7 cents each day, how many days before all his money will be spent? 35÷7=how many? Why?

4. If Henry has 56 cents, how many oranges can he buy at 7 cents each? 56÷7=how many? Why?

---

* The purpose in this exercise is to show how we may find how many times one number is contained in another by taking the former from the latter as many times as possible; i. e., by subtraction. See note at the bottom of page 108. That process may be made serviceable for this purpose.

Copy, fill out, and learn the following :

$6 \div 6 =$	$7 \div 7 =$	$36 \div 6 =$	$6 \div 6 =$
$12 \div 6 =$	$14 \div 7 =$	$42 \div 7 =$	$7 \div 7 =$
$18 \div 6 =$	$21 \div 7 =$	$42 \div 6 =$	$14 \div 7 =$
$24 \div 6 =$	$28 \div 7 =$	$30 \div 6 =$	$18 \div 6 =$
$30 \div 6 =$	$35 \div 7 =$	$35 \div 7 =$	$21 \div 7 =$
$36 \div 6 =$	$42 \div 7 =$	$49 \div 7 =$	$54 \div 6 =$
$42 \div 6 =$	$49 \div 7 =$	$24 \div 6 =$	$56 \div 7 =$
$48 \div 6 =$	$56 \div 7 =$	$12 \div 6 =$	$48 \div 6 =$
$54 \div 6 =$	$63 \div 7 =$	$70 \div 7 =$	$63 \div 7 =$
$60 \div 6 =$	$70 \div 7 =$	$60 \div 6 =$	$28 \div 7 =$

Fourth Exercise.

1. Make 18 a's on your slate thus :

a a a a a a a a a a a a a a a a a a

Then mark them off into groups of 9 each. How many such groups will you have? How many 9's in 18? How many times 9 are 18? $18 \div 9 =$ how many?

2. Make 36 a's on your slate thus :

a a a a a a a a a a a a a a a a a a
a a a a a a a a a a a a a a a a a a

Then make 9 large circles thus:

Then rub out one *a* and put it in one of the circles. Then rub out another *a* and put it in another circle. Then another and another, till you have one *a* in each circle. Then go round again and

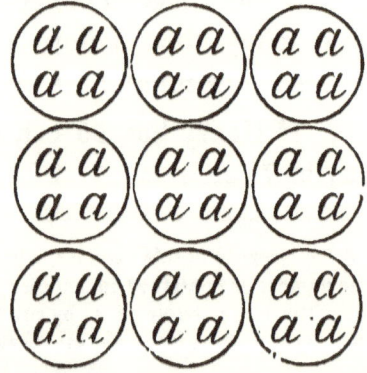

put another *a* in each of the circles, till you have 2 *a*'s in each circle. Then go round again, putting another *a* in each circle, till all the 36 *a*'s are used up. How many *a*'s will there be in each circle? If you divide 36 into 9 equal groups, how many are there in each group? $36 \div 9 =$ how many? Why?

5. Copy, fill out, and learn the following:

$8 \div 8 =$	$9 \div 9 =$	$72 \div 9 =$	$27 \div 9 =$
$16 \div 8 =$	$18 \div 9 =$	$72 \div 8 =$	$32 \div 8 =$
$24 \div 8 =$	$27 \div 9 =$	$8 \div 8 =$	$36 \div 9 =$
$32 \div 8 =$	$36 \div 9 =$	$9 \div 9 =$	$40 \div 8 =$
$40 \div 8 =$	$45 \div 9 =$	$80 \div 8 =$	$45 \div 9 =$
$48 \div 8 =$	$54 \div 9 =$	$90 \div 9 =$	$56 \div 8 =$
$56 \div 8 =$	$63 \div 9 =$	$16 \div 8 =$	$63 \div 9 =$
$64 \div 8 =$	$72 \div 9 =$	$18 \div 9 =$	$64 \div 8 =$
$72 \div 8 =$	$81 \div 9 =$	$24 \div 8 =$	$81 \div 9 =$
$80 \div 8 =$	$90 \div 9 =$	$48 \div 8 =$	$54 \div 9 =$

Fifth Exercise.

1. How many 10's are there in 20? How many in 30? In 40?

2. $40 \div 10 =$ how many? $30 \div 10 =$ how many?

3. How many 1's in 6? $6 \div 1 =$ how many? How many 1's in 7? $7 \div 1 =$ how many?

4. Mary has 80 needles in 8 papers, with the same number in each paper. How many needles in each paper?

5. If Mary puts her 80 needles up in papers of 10 needles each, how many papers will she have? $80 \div 8 =$ how many? $80 \div 10 =$ how many?

6. John has 60 cents in his bank. If he takes out 10 cents each day, how many days before his money will be gone? How many times can you take 10 out of 60? $60 \div 10 =$ how many?

7. Copy, fill out, and learn the following:

$10 \div 10 =$	$1 \div 1 =$	$50 \div 10 =$	$3 \div 1 =$
$20 \div 10 =$	$2 \div 1 =$	$8 \div 1 =$	$2 \div 1 =$
$30 \div 10 =$	$3 \div 1 =$	$30 \div 10 =$	$60 \div 10 =$
$40 \div 10 =$	$4 \div 1 =$	$70 \div 10 =$	$80 \div 10 =$
$50 \div 10 =$	$5 \div 1 =$	$7 \div 1 =$	$5 \div 1 =$
$60 \div 10 =$	$6 \div 1 =$	$4 \div 1 =$	$90 \div 10 =$
$70 \div 10 =$	$7 \div 1 =$	$10 \div 10 =$	$6 \div 1 =$
$80 \div 10 =$	$8 \div 1 =$	$1 \div 1 =$	$9 \div 1 =$
$90 \div 10 =$	$9 \div 1 =$	$20 \div 10 =$	$100 \div 10 =$
$100 \div 10 =$	$10 \div 1 =$	$40 \div 10 =$	$10 \div 1 =$

Sixth Exercise.

1. Here is a beautiful pansy. How many blossoms are there on it? If you make 2 bouquets and put 3 of these pansies in each of them, how many pansies will remain? How many 3's in 8, and how many over?

2. John had 15 cents and gave 4 cents apiece for 3 oranges. How many cents did he have remaining? How many times can you take 4 from 15? How many will remain after you have taken 4 from 15 as many times as you can?

3. Are there 4 2's in 7? How many 2's are there in 7? Is there any remainder after you have taken 3 2's from 7? What is it?

4. How many are 5 times 7? 6 times 7? How many 7's are there in 38, and how many over? How many 7's in 40, and how many over? How many 7's in 37, and how many remaining?

5. If you divide 12 into as many 5's as you can, how many 5's will you have, and how many remaining? If you divide 23 into as many 6's as you can, what will the remainder be? $3 \times 6 =$ how many? $4 \times 6 =$ how many? Are there 4 6's in 23?

6. Say the "3 times" of the Multiplication Table. Are there 2 3's in 17? Are there 3 3's in 17? Are there 4? Are there 5? Are there 6? How many 3's are there in 17, and how many over?

7. Say the "6 times" of the Multiplication Table. Are there 3 6's in 27? Are there 4? Are there 5? If you divide 27 into as many 6's as you can, how many 6's will you have, and what remainder?

8. Say the "8 times" of the Multiplication Table. Are there 3 8's in 47? Are there 4? Are there 5? Are there 6? How many times is 8 contained in 47, and how many remain?

Seventh Exercise.

1. Say the "4 times" of the Multiplication Table. Are there 3 4's in 27? Are there 5 4's in 27? Are there 6? Are there 7? How many 4's in 27, and what is the remainder? 27÷4 are how many, and what is the remainder?

2. John had 35 cents and bought 8 lead-pencils, at 4 cents each. How much money had he left? How many 4's in 35, and how many remain?

3. Copy, and fill out the following:

14÷3 = —— and —— remainder.
22÷4 = —— and —— remainder.
18÷4 = —— and —— remainder.
37÷5 = —— and —— remainder.
40÷6 = —— and —— remainder.
51÷7 = —— and —— remainder.
78÷9 = —— and —— remainder.
67÷8 = —— and —— remainder.
80÷9 = —— and —— remainder.

Drill Exercise.*

1. Add 2, 3, 4, subtract 6, multiply by 7, divide by 3, subtract 5, add 7, divide by 3, multiply by 8, divide by 6, add 9, add 7, divide by 5, divide by 2, add 8, 3, 9, 2, 4, divide by 7. What is the result?

---

* We repeat that this character of exercise, either wholly oral or by means of blackboard, arithmetical roll, or lattice, must be constantly kept up. No day should pass in a primary school without more or less of this drill in combining numbers.

2. Add 5 to 7, subtract 8, multiply by 3, add 9, divide by 7, add 8, subtract 9, add 6, multiply by 7, add 8, divide by 8. What is the result?

3. From 13 subtract 8, multiply by 4, add 6, add 4, add 2, divide by 4, divide by 2, multiply by 7, add 8, divide by 9. What is the result?

4. Divide 54 by 6, divide by 3, multiply by 9, add 8, add 10, divide by 9, multiply by 6, add 5, divide by 7, multiply by 2, multiply by 7, add 2, divide by 8, multiply by 6, add 9, divide by 7, divide by 9, subtract 1. What is the result?

5. Divide 27 by 9, multiply by 3, add 5, add 4, divide by 6, multiply by 8, add 8, add 4, divide by 6, multiply by 7, add 7, divide by 7, multiply by 9, add 1, divide by 8. What is the result?

6. Divide 56 by 7, multiply by 6, add 6, divide by 9, multiply by 7, add 7, 3, 6, 4, 1, divide by 9, multiply by 8, add 10, 10, 5, divide by 9. What is the result?

7. From 15 subtract 9, add 2, multiply by 8, add 6, divide by 10, multiply by 9, add 1, divide by 8, subtract 1, multiply by 7, add 7, divide by 9. What is the result?

8. Divide 56 by 7, divide by 2, multiply by 8, add 3, divide by 7, add 4, multiply by 6, add 9, divide by 7. What is the result?

9. From 13 subtract 8, multiply by 6, add 6, divide by 9, multiply by 8, add 4, divide by 6, subtract 6, multiply by 3. What is the result?

6

SECTION III.

FRACTIONS.

Purpose.—*To teach the signification of the fractions $\frac{1}{2}$, $\frac{2}{3}$, $\frac{1}{3}$, $\frac{2}{3}$, $\frac{1}{4}$, $\frac{1}{4}$, $\frac{2}{4}$, $\frac{3}{4}$, $\frac{1}{5}$, $\frac{1}{6}$, $\frac{2}{5}$, etc., to tenths inclusive, and how to get any such fractional part of a number.*

First Exercise.

1. If you cut an apple into 2 parts of just the same size, what do you call one of the parts? If you cut an apple into 2 parts so that one of the parts shall be larger than the other, will either of the parts be a half an apple? In the lower picture, is the part of the apple on the right more than half the apple, or less than half?

2. When you divide anything into halves, how many parts do you make of it? How many halves in the whole of anything? Which half of an orange is the larger?

3. Into how many parts is this apple divided? Are the parts of equal size? What is one of the 3 equal parts of anything called? *Ans.* One Third.

4. Into how many parts is this apple divided? Is this apple divided into thirds? Why not?

5. How many thirds are there in the whole of anything? Which third of an orange is the largest? If you divide an

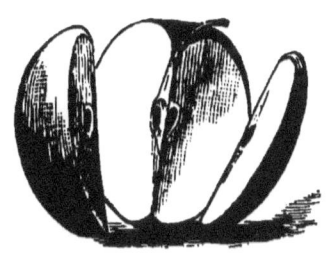

apple into thirds and give away two of them, how many will you have left? What part of the apple will you have left? What part will you have given away?

6. Here are 10 cherries on a plate. If you divide them equally between two girls, what part of the cherries will each girl have? How many will each girl have? One-half of 10 is how many? $10 \div 2 =$ how many? By what must you divide a number to get one-half of it?

7. Here are 12 nuts. If you divide them equally among 3 boys, what part of the nuts will each boy receive? How many nuts will each boy have? $12 \div 3 =$ how many? By what must we divide a number to get one-third of it?

8. One-half is written $\frac{1}{2}$.
One-third is written $\frac{1}{3}$.
Two-thirds is written $\frac{2}{3}$.

What does $\frac{1}{3}$ mean? What does $\frac{1}{2}$ mean? What does $\frac{2}{3}$ mean?

9. Copy and fill out the following:

$\frac{1}{2}$ of 8=	$\frac{1}{2}$ of 18=	$\frac{2}{3}$ of 6=	$\frac{2}{3}$ of 3=
$\frac{1}{3}$ of 9=	$\frac{1}{3}$ of 18=	$\frac{2}{3}$ of 12=	$\frac{2}{3}$ of 15=
$\frac{1}{2}$ of 6=	$\frac{1}{2}$ of 16=	$\frac{2}{3}$ of 9=	$\frac{2}{3}$ of 18=
$\frac{1}{3}$ of 6=	$\frac{1}{3}$ of 21=	$\frac{2}{3}$ of 21=	$\frac{1}{2}$ of 1=
$\frac{1}{2}$ of 10=	$\frac{1}{2}$ of 2=	$\frac{2}{3}$ of 30=	$\frac{1}{3}$ of 1=
$\frac{1}{2}$ of 4=	$\frac{1}{3}$ of 3=	$\frac{2}{3}$ of 27=	$\frac{2}{3}$ of 1=

Second Exercise.

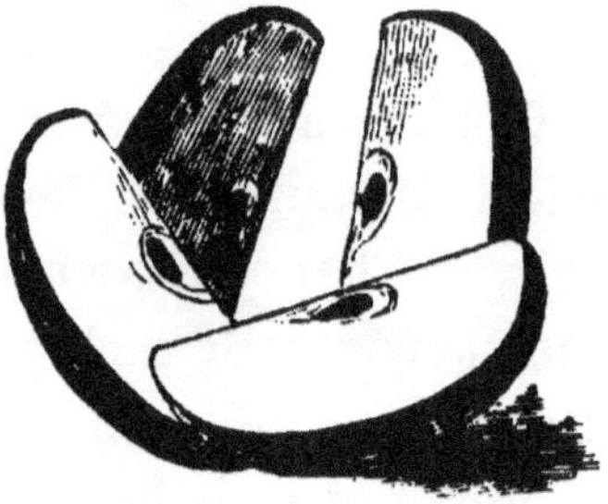

1. Into how many equal parts is this apple divided? If anything is divided into four equal parts, what are the parts called? *Ans.* Fourths. How many fourths in the whole of anything?

2. If John has a cake and gives one-fourth of it to Henry, one-fourth to Mary, and one-fourth to Jane, how much has he left?

3. If you divide 12 flowers equally among four boys, what part of them all does one boy get? How many flowers does one boy get? One-fourth of 12 is how many? $12 \div 4 =$ how many? By what do you divide to get one-fourth of any number?

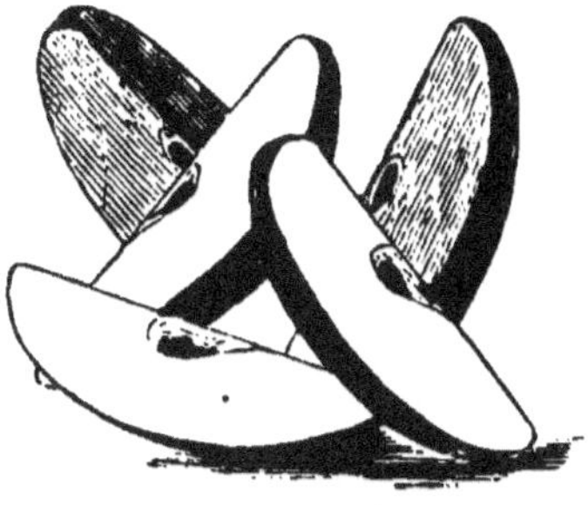

4. Into how many parts is this apple divided? Are the parts equal? If anything is divided into 5 equal parts, what is any one of the parts called? *Ans.*, Fifths. How many fifths in the whole of anything?

5. If James gives away three-fifths of his melon, how many fifths will he have left? Into how many parts must a melon be divided so that the parts shall be fifths? How many fifths in a whole melon?

6. Henry has 15 nuts, and divides them equally between 5 boys. What part of all the nuts does one boy receive? If 2 of the boys put their shares together, what part of all the nuts will they make? One-fifth and one-fifth make how many fifths?

7. How many nuts are one-fifth of 20 nuts? $20 \div 5 =$ how many?

8. How many are two-fifths of 20 nuts? Three-fifths? By what do you divide to get one-fifth of any number?

9. One-fourth is written $\frac{1}{4}$. One-fifth is written $\frac{1}{5}$. Two-fourths is written $\frac{2}{4}$. Two-fifths is written $\frac{2}{5}$. The number above the short line shows how many fourths or fifths are meant.

10. What does $\frac{3}{5}$ mean? What does $\frac{3}{4}$ mean? What $\frac{4}{5}$?

11. Copy, and fill out the following:

$\frac{1}{4}$ of 12=	$\frac{3}{4}$ of 20=	$\frac{3}{5}$ of 35=	$\frac{2}{5}$ of 30=
$\frac{1}{5}$ of 10=	$\frac{3}{4}$ of 12=	$\frac{4}{5}$ of 45=	$\frac{3}{5}$ of 30=
$\frac{1}{5}$ of 30=	$\frac{3}{4}$ of 32=	$\frac{2}{5}$ of 25=	$\frac{4}{5}$ of 30=
$\frac{1}{5}$ of 40=	$\frac{3}{4}$ of 10=	$\frac{2}{5}$ of 10=	$\frac{5}{5}$ of 30=
$\frac{1}{4}$ of 32=	$\frac{4}{5}$ of 35=	$\frac{2}{5}$ of 15=	$\frac{4}{4}$ of 8=

Third Exercise.

1. Into how many parts is this apple divided? Are the parts equal? One of the 6 equal parts into which the whole of anything may be divided is called a sixth. How many sixths in the whole of anything?

2. One of the six equal parts of the apple is called what? Then 2 of the six equal parts would be what? Three? Four? Five?

3. You see that one-fourth of the whole of anything is one of the 4 equal parts of it. One-fifth is one of 5 equal parts. One-sixth is one of 6 equal parts. What, then, is one-seventh of the whole of anything? One-eighth? One-ninth? One-tenth?

4. If an apple is divided into 7 equal parts, what is one of the parts called? What are 2 of the 7 equal parts called? Three? Four? Five? Six? Seven?

5. Here is the whole of an apple which has been divided into 8 equal pieces. Part of the pieces are on one plate, and part are on the other. How much of the apple is on the upper plate? How much on the lower?

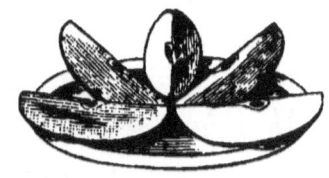

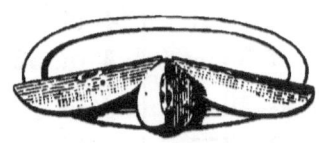

6. If you have 36 nuts and divide them equally among 9 boys, what part of the whole does one boy get? How many nuts does one boy get? How many do 2 boys get? Three boys? Three-ninths of 36 are how many? How do you get one-ninth of any number?

7. What does $\frac{1}{4}$ mean? $\frac{1}{3}$? $\frac{1}{5}$? $\frac{1}{7}$? $\frac{1}{8}$? $\frac{1}{9}$? $\frac{1}{10}$?

8. What does $\frac{2}{3}$ mean? $\frac{3}{5}$? $\frac{2}{5}$? $\frac{3}{8}$? $\frac{4}{8}$? $\frac{5}{8}$? $\frac{4}{7}$? $\frac{3}{7}$? $\frac{4}{5}$? $\frac{4}{9}$? $\frac{3}{8}$? $\frac{5}{8}$? $\frac{4}{8}$? $\frac{7}{8}$? $\frac{6}{8}$? $\frac{5}{8}$? $\frac{8}{8}$? $\frac{7}{10}$? $\frac{5}{10}$? $\frac{7}{10}$? $\frac{8}{10}$?

Fourth Exercise.*

1. Copy, and fill out the following:

$\frac{2}{3}$ of 12=	$\frac{3}{5}$ of 50=	$\frac{3}{4}$ of 56=
$\frac{2}{5}$ of 10=	$\frac{4}{7}$ of 21=	$\frac{5}{9}$ of 81=
$\frac{4}{7}$ of 14=	$\frac{5}{6}$ of 54=	$\frac{5}{8}$ of 72=
$\frac{5}{9}$ of 18=	$\frac{4}{7}$ of 63=	$\frac{8}{10}$ of 90=
$\frac{3}{4}$ of 12=	$\frac{4}{9}$ of 81=	$\frac{8}{10}$ of 90=
$\frac{3}{8}$ of 16=	$\frac{7}{8}$ of 56=	$\frac{7}{10}$ of 80=
$\frac{3}{10}$ of 40=	$\frac{5}{8}$ of 40=	$\frac{5}{8}$ of 72=
$\frac{5}{6}$ of 30=	$\frac{3}{4}$ of 24=	$\frac{5}{8}$ of 64=
$\frac{4}{7}$ of 35=	$\frac{4}{8}$ of 32=	$\frac{7}{9}$ of 72=
$\frac{5}{6}$ of 42=	$\frac{5}{8}$ of 72=	$\frac{6}{10}$ of 70=

2. Write in figures on your slate, one-half, one-third, two-thirds, three-fourths, five-eighths, three-eighths, 5 sixths, 4 sevenths, eight-ninths, 5-ninths, three-tenths, 7-tenths, 4-ninths, 2-ninths, 2-fifths, two-sevenths.

---

* Exercises of this character should be assigned by writing them on the blackboard until they can be performed with the utmost ease. It affords an excellent drill in division and multiplication.

Fifth Exercise.*

1. Here are how many whole apples? How many half apples? How many in all? We write Three and a half thus: $3\frac{1}{2}$.

2. How many whole apples are there in this picture? How many pieces? What are the pieces—halves, thirds, or quarters? The number of apples in this picture is written thus: $2\frac{2}{3}$. Can you tell what $2\frac{2}{3}$ means?

3. Read the following: $1\frac{1}{2}$, $2\frac{1}{2}$, $4\frac{1}{3}$, $5\frac{2}{3}$, $3\frac{7}{8}$, $6\frac{8}{9}$, $10\frac{5}{7}$.

4. There are 5 apples on the plate. If John takes half and Henry half, how many will each have? Write the number.

5. If 7 apples are divided equally among 2 boys, how many will each boy have? If each boy takes 3, how many will remain? What must they do with that? What does $3\frac{1}{2}$ mean?

6. If 15 apples are to be divided equally between 4 boys,

---

* The purpose of this exercise is to teach the meaning of such mixed numbers as $4\frac{1}{2}$, $3\frac{2}{3}$, $10\frac{3}{8}$, $1\frac{3}{4}$, etc., and how to read them.

how many will each boy have ? If each boy takes 3 apples, how many will be left? If now these 3 apples which are left be divided into fourths, how many fourths will they make ? Now if these 12 pieces are divided equally among the 4 boys, how many of them will each boy get ? How many whole apples will each boy have? How many fourths? What is $4\frac{3}{4}$?

7. If 14 apples are divided equally among 3 boys, how many whole apples and how many thirds will each boy get? $14 \div 3 =$ how many, and how many over?

8. If 30 apples are divided equally among 7 boys, how many whole apples will each boy have? $30 \div 7 =$ how many, and what remainder? After each boy has received his 4 whole apples, how many apples are there left? In order to divide an apple equally among 7 boys, into what parts must it be divided? If each of 2 apples is divided into 7 parts, how many parts are there? How many whole apples and how many sevenths will each boy have?

9. If 35 oranges are divided equally among 8 boys, how many will each boy have? Will he have $4\frac{5}{8}$ or $4\frac{3}{8}$?

Practical Exercises.

1. John and James bought a melon worth 8 cents, which they are to share equally. How much ought each to pay? What part of the price must John pay? $\frac{1}{2}$ of 8 is how much ? How do you get $\frac{1}{2}$ of any number ?

2. John, James, and Henry bought a pie worth 12 cents, which they are to share equally. How much must each pay? What part of the price must Henry pay? How

do you get $\frac{1}{3}$ of any number? How much must John and James together pay? What part of the pie do John and James together own? $\frac{2}{3}$ of 12 = how much?

3. Mary and Jane bought a doll for 63 cents. Mary paid $\frac{5}{9}$ of the price. What part of the price did Jane pay? $\frac{5}{9}$ of 63 = how much? $\frac{4}{9}$ of 63 = how much? How many cents did Mary pay? How many did Jane pay?

4. Henry started to market with 56 eggs, but broke $\frac{1}{7}$ of them. How many did he break? What part of the eggs remained unbroken? How many eggs were unbroken?

5. Mary's hen had a brood of 12 chickens, but a hawk caught $\frac{2}{3}$ of them. How many had she left? If $\frac{2}{3}$ were caught, how many thirds remained? $\frac{1}{3}$ of 12 = how many?

6. John's suit—coat, vest, and pantaloons—cost 9 dollars. The coat and vest cost $\frac{2}{3}$ of the whole. What part of the whole did the pantaloons cost? How many dollars did the pantaloons cost?

7. Mary has $\frac{3}{8}$ of an apple, Jane $\frac{2}{8}$, and Henry the rest. How much of the apple has Henry?

8. If one cord of wood costs 6 dollars, how much will $\frac{1}{3}$ of a cord cost? $\frac{1}{2}$ of a cord? $\frac{2}{3}$ of a cord?

9. I had 21 dollars, and spent $\frac{2}{7}$ of it for a pair of boots, and $\frac{3}{7}$ for a coat. What part of my money had I left? How many dollars?

10. Our cistern was entirely dry on Saturday. But it rained on Sunday, and filled it $\frac{1}{5}$ full. On Monday it rained again, and the cistern filled up so that it was $\frac{4}{5}$ full. How much ran in on Monday?

SECTION IV.
DENOMINATE NUMBERS.

Purpose.—*To teach a few of the more common denominations of measure, weight, and money, so that the pupil shall have a clear conception of each, and a knowledge of their mutual relations.*

First Exercise.

UNITED STATES MONEY.

1. What is this a picture of? What is such a piece of money made of?

Answer.—This is a picture of our common cent, which is made of bronze. Bronze is copper and tin melted together.

2. What is this a picture of? What is such a piece of money made of? How many cents is such a piece worth?

3. How many cents is a half dime? How many cents in 2 dimes? In 1½ dimes?

4. Here are two pieces of money. What are they? Which is the most? If they are both  dollars, why is it that one is so much larger than the other? Will the little one buy just as much as the big one? How many cents is a dollar worth? How many dimes in 100 cents? $100 \div 10 =$ how many? How many dimes make a dollar?

5. What piece of money is this? What is it made of? How many  dimes in a dollar? Then how many dimes is a half-dollar worth? Five dimes are how many cents? How many cents in a half-dollar?

6. What are these pieces? Which is worth the most? Why is one so much larger  than the other? How many dimes in a half-dollar? Then how many 5-cent nickels does it take to make half a dollar? How many half-dimes make a half-dollar?

7. Here is a quarter of a dollar. How many quarters of a dollar does it take to make a half-dollar? What part of  a half-dollar is a quarter-dollar? How many 5-cent nickels does it take for half a dollar? Then how many for

a quarter of a dollar? How many cents in a half-dime? How many half-dimes in a quarter-dollar? How many cents in a quarter-dollar?

8. What is a ten-cent piece called? What is a 50-cent piece called? What is a 25-cent piece called?

9. Learn this

TABLE OF UNITED STATES MONEY.

10 *cents* $= 1$ *dime.*
10 *dimes* $= 1$ *dollar.*
25 *cents* $= \frac{1}{4}$ *dollar.*
50 *cents* $= \frac{1}{2}$ *dollar.*

9. John bought an arithmetic for half a dollar, a slate for 2 dimes, a sponge for half a dime, and a pencil for a cent. How much did all cost him?

10. The character $ signifies dollars, and is written before the figure or figures telling how many. Thus, $8, means 8 dollars. $23 means 23 dollars, etc.

11. Figures representing cents are written right after those representing dollars, with a period, called a *Decimal Point,* between the dollars and cents. Thus, $12.15 means 12 dollars and 15 cents. $58.37 means 58 dollars and 37 cents. *c.,* or *ct.,* is used as an abbreviation for cents. Thus, 28 *c.,* or 28 *ct.,* is 28 cents.

12. Read $62.25; $5.18; $7.30; $19.03. The last is 19 dollars and 3 cents, since 03 is just the same as 3. We have to put the 0 before the 3 when we write dollars and cents together; otherwise we could not tell whether the 3 did not mean 30 cents. Thus, $19.3 would be the same as $19.30.

13. Read $8.05; $8.50; $10.10; $10.01; $0.58; $0.23; $100; $100.05.

Second Exercise.*

MEASURES OF LENGTH.†

1. When we wish to tell how long anything is, we say it is so many inches, feet, yards, rods, or miles. You will need to learn just how long each of these measures is. This short line is 1 inch long, and the long one is 3 inches long.

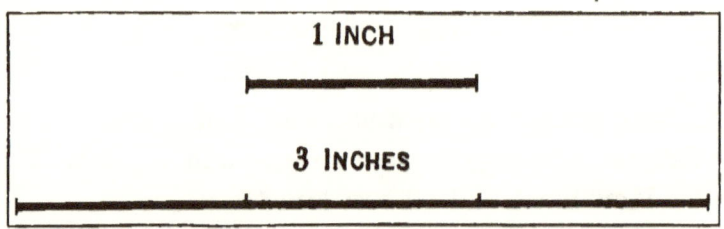

2. Get a little stick, or a little narrow slip of paper, and cut it off just 1 inch long; that is, just as long as the short mark. Then make a mark on your slate 4 times as long as your inch measure. Make another 5 times as long. Make another 6 inches long. Another 7 inches long.

---

* The exercises following these tables are only given as specimens of what the teacher should do in connection with the memorizing of the tables by the pupils. It is specially true on this subject that no book can supply what the pupil needs. He must learn to judge of measures and weights—*i. e.*, to have some just conception of magnitudes and quantities, and to measure them by the proper apparatus, as measuring-rulers, rods, cups, weights, etc.

† No good work can be done in this subject without a little apparatus. Thus, for length, a foot-ruler, divided into inches, and a yard-stick, divided on one side into feet and on the other into halves, quarters, and eighths, as on the dry-goods merchant's counter. These the pupils must handle and apply. Practice in guessing, and then testing the guess, will be entertaining and profitable. The apparatus needed for other uses will be specified in its place.

3. Is this page 5 inches wide? Is it 4 inches wide? Is it any more than 4 inches wide? How long is this page?

4. Get another stick long enough so that you can cut off a piece just 12 times as long as your inch measure.* A stick that is 12 inches long is just a foot long. So we say, 12 inches make 1 foot. Or we write it, 12 inches = 1 foot.

5. How many inches long is your desk? Do you think it is a foot long? Is it a foot wide?

6. Learn the following

TABLE OF LINEAR MEASURES.

12   *inches* = 1 *foot.*
3   *feet*   = 1 *yard.*
5½ *yards* = 1 *rod.*
320   *rods*   = 1 *mile.*

Abbreviations.—in. stands for inch, or inches; *ft.* for foot, or feet; *yd.* for yard, or yards; *rd.* for rod, or rods; and *mi.* for mile, or miles.

7. Do you think the door is a yard wide? How many yards high do you think it is?

8. If a door is two *yds.* high, how many *feet* high is it? How many feet long is a blackboard that is 3 *yds.* long?

9. How many yards make a rod? Do you think this room is a rod wide? Is it more than a rod wide? How many rods long do you think it is?

---

* The teacher should allow (*require*, if necessary) each pupil to have such sticks, and measure with them.

10. How many rods wide do you think the school-yard is? If you had a stick 1 yard long, could you measure and find out how wide the yard is? How many times the length of the yard-stick does it take to make a rod?

11. How many feet long is a yard? How many feet is a half a yard? How many inches in a foot? How many inches in one foot and a half? How many inches in half a yard?

12. This ruler is divided as a yard-stick is usually, only it is but 3 *in.* long instead of 3 *ft.* What part of the

whole length is it from either end to the two dots? What part from either end to the one dot nearest that end? What part is it from the one dot to the two dots? How many fourths of a yard in a half-yard?

13. Do you know what other name we give to one-fourth of anything? We often call it a quarter. How many inches in a half a yard? How many in a quarter of a yard?

14. If there are 9 *in.* in a quarter of a yard, and 4 quarters in a yard, how many inches are there in a yard?

15. How many inches in a half a foot? How many in a quarter of a foot?

16. How many inches in $1\frac{1}{3}$ *ft.*? In $1\frac{1}{4}$ *ft.*?

Note.—The teacher should give the pupils as good an idea of a mile as possible, by referring to distances with which they are familiar, as to some house a mile off, a half-mile off, 2 miles off, etc. Also by the time it takes them to walk a mile, etc. Such questions as this will help: Would it tire you to walk a rod? Two rods? A mile? Two miles?

Third Exercise.*

MEASURES FOR LIQUIDS.

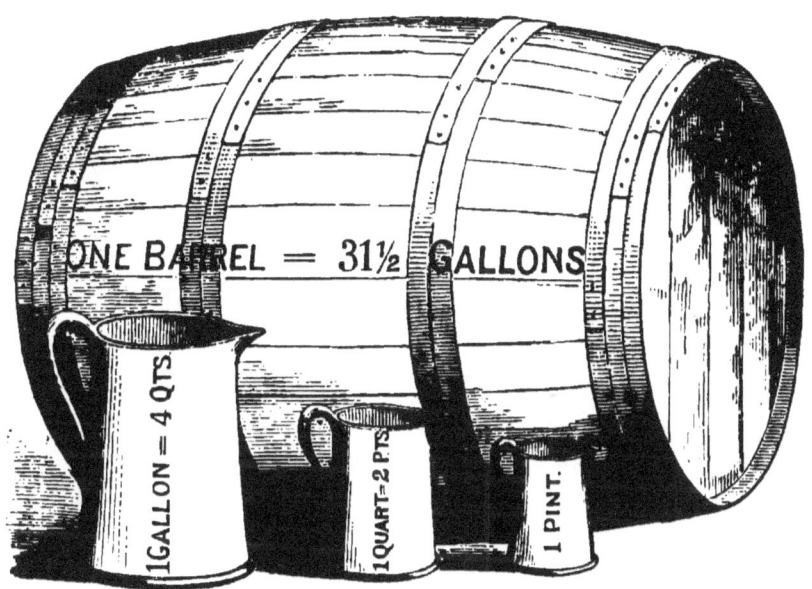

1. If you were to go to the grocery to buy molasses, or kerosene, or vinegar, how would you tell the groceryman how much you wanted? Which is the more, a pint or a quart? Which is the more, a quart or a gallon? Do you know about how large a cup it takes to hold a pint? A pint cup will hold about twice as much as a common tea-cup.

---

* To teach this subject properly, a gallon measure, quart measure, and pint measure are essential. Fill the quart measure from the pint, and *vice versa*. Also the gallon from the quart, etc. Half-gallons and half-pints are useful also.

2. Learn the following

TABLE OF LIQUID MEASURES.

2 *pints* = 1 *quart.*
4 *quarts* = 1 *gallon.*
31½ *gallons* = 1 *barrel.*

Abbreviations.—pt. stands for pint, or pints; *qt.* for quart, or quarts; *gal.* for gallon, or gallons; and *bbl.* for barrel, or barrels.

3. Do you think a common water-pail holds a gallon? Do you think it holds 2 *gal.*? 3 *gal.*? 4 *gal.*?

4. Do you think that a common drinking tumbler holds a quart? Do you think it holds a pint? How much do you think it holds?

5. I have heard a boy who was very thirsty say that he could drink a gallon. Could he? Could he drink a pint? A quart?*

6. If you wanted to measure out a gallon of water and had nothing but a pint cup to measure with, how many cupfuls would you have to take?

7. What part of a quart is a pint? What part of a gallon is a quart?

8. How many pints in 7 *qt.*?

9. How many quarts in 10 *pt.*?

10. How many quarts in 10 *gal.*?

11. How many pints in 3 *gal.*?

---

* In such ways and by allowing (requiring) the pupils to use the measures, seek to give them correct notions of these measures. Let them find out by actual trial that 2 pints = 1 quart, and that 4 quarts = 1 gallon.

Fourth Exercise.*

MEASURES FOR GRAINS, SEEDS, ETC.

1. Here are two cups, and each is called a quart cup. Are they of the same size? Measure them and see which is the wider. Which is the higher? Well, the smaller one is such a quart cup as we measure milk, water, vinegar, or any liquid in; while the larger is such a quart cup as we use to measure seeds, grain, and any *dry* substances which we wish to measure in this way. So you see that a quart of wheat is more than a quart of water. So also a pint of corn is more than a pint of milk. It takes about 7 quarts of liquid measure to make as much as 6 quarts of dry measure; † because the quart cup by which we measure liquids is so much smaller than that by which we measure grain, seeds, and other dry substances.

2. Learn the following

TABLE OF DRY MEASURES.

2 *pints* = 1 *quart*.
8 *quarts* = 1 *peck*.
4 *pecks* = 1 *bushel*.

---

* For teaching the *Dry Measures*, a pint measure, a quart measure, a four-quart measure, peck measure, and half-bushel measure are important. A bushel basket would also be well.

† By all means have the pupils see this and all kindred facts exemplified with the measures themselves.

Abbreviations.—pk. stands for peck, or pecks; and *bu.* for bushel, or bushels.

3. Do you think a common wooden water-pail will hold a bushel ? Will it hold a half-bushel ? Will it hold a peck ? *

4. How many common wooden water-pailfuls of corn do you think it would take to fill a bushel basket?

5. Do you think you could carry a peck of corn ? Could you carry a bushel ? Two bushels?

6. You have seen flour-barrels, have you not? How many bushels do you think a flour-barrel holds? *2 bu. ?* *3 bu. ?* *4 bu. ?*†

7. Do you think that a boy can put a peck of nuts in his pocket? Can he put a bushel of nuts in his pocket? Can he put a quart in ? A pint?

8. How many pints does it take to make a peck ?

9. How many pecks in 5 *bu.* ?

10. How many half-bushel measures full does it take to fill a two-bushel bag?

11. If I want to measure out a bushel of wheat and have only a quart cup to do it with, how many cupfuls must I take?

12. If I wish to measure out 5½ *bu.* of corn, how many times must I fill the half-bushel measure?

13. What part of a quart is a pint?

14. What part of a peck is a quart?

15. What part of a peck is 2 *qt.?* Three quarts? 4 *qt.?* 5 *qt.?* 6 *qt.?* 7 *qt.?* 8 *qt.?*

---

* Such a pail holds about 10 liquid quarts, or about a pint over a peck.

† Such a barrel is 27 inches deep and about 18 inches in diameter, and hence holds about 104 dry quarts.

16. What part of a bushel is a peck? *2 pk.*? *3 pk.*?

17. How many pecks in a half-bushel?

18. How many times will you have to fill the 4-quart measure to make a half-bushel? How many times to make a bushel?

Fifth Exercise.

WEIGHTS AND WEIGHING.

1. If you were to go to the grocery to buy some tea, coffee, or sugar, how would you tell the groceryman how much you wanted? Would you tell him that you wanted a pint of tea, or a yard of coffee, or a gallon of sugar? How would you tell him? Would a pound of sugar fill a gallon measure? Which measure do you think a pound of sugar would come nearest to filling—a pint, quart, or gallon measure? What would the groceryman use to determine how much sugar he gave you? (Scales.)

2. Learn this

TABLE OF AVOIRDUPOIS* WEIGHTS.

16 *ounces*	= 1 *pound.*
100 *pounds*	= 1 *hundred-weight.*
20 *hundred-weight*	= 1 *ton.*

---

* If the teacher thinks best, she can explain that this long word is three French words (*avoir du poids*) put together, and means *to have weight.*

Abbreviations.—oz. stands for ounce, or ounces; *lb.* for pound, or pounds; *cwt.* for hundred-weight; and *T.* for ton, or tons.

3. How many 5-cent nickels do you think it takes to make an ounce?* How many to make a pound?

4. How much do you think a pint of water weighs?† It weighs just about a pound. How many ounces, then, does a common teacupful of water weigh? What part of a pound?

5. How much does a quart of water weigh?‡ How much does a gallon of water weigh? How much does a common wooden pailful of water weigh? You remember that we learned that such a pail holds about 10 quarts.

6. How much do you weigh? Do you weigh a ton? A hundred-weight? How many boys who weigh 50 *lb.* each does it take to weigh a hundred-weight?§ How many to weigh a ton?

7. Did you ever see a large load of hay drawn by two horses? Do you think such a load weighs a hundred-weight? Do you think a span of horses could draw a ton of hay? Can a span of horses draw a ton of boys and girls?

8. How many ounces in a pound? How many in a half-pound? How many in a quarter of a pound?

9. Four ounces is what part of a pound? 8 oz. is what part of a pound? 2 oz. is what part of a pound?

---

* An avoirdupois ounce = 437.5 grains, and a 5-cent nickel weighs 77.16 grs.

† A quart weighs 2.0843 + lbs. A pint of water, therefore, is an excellent object with which to teach the pupil what a pound is.

‡ Teach them to say *nearly* in such cases; also that the vessel is not included.

§ This is beyond what the pupil has been taught; but it affords so good an illustration, that it will be well for the teacher to explain it; *first, however, letting the class try their full strength on it.* Very likely they can get it out.

10. How many 2-*oz.* weights would it take to make a pound? How many 4-*oz.* weights? How many 8-*oz.* weights?

11. How many ounces in ¾ *lb.*? How many in 1½ *lb.*?

12. Four ounces and 5 ounces and six ounces together lack how many ounces of being a pound?

13. How many hundred-weight make a ton? How many make a quarter of a ton? How many make a half ton? 40 *cwt.* are how many tons?

Sixth Exercise.

WEIGHING WITH BALANCE.*

1. Here is a *Balance.* It is the simplest machine used for weighing? All you have to do is to put into one pan such weights as are equal to the amount you want to weigh, and pour the thing to be weighed into the other pan till the pans balance.

2. If the man in the picture has a 2-pound weight and

---

* The purpose of this and the two following exercises is to teach the pupils how to weigh. Every primary school should have a balance with weights, a pair of steel-yards, and a set of grocer's scales, and the pupils should be taught to use them.

a 4-pound weight in one pan, how much coffee will he have in the other to make them balance?

3. I wanted to find out how much a dressed chicken weighed, and put it into one pan, and then put into the other pan a 2-pound weight, and a 1-pound weight, and an 8-*oz.* weight, and a 4-*oz.* weight. How much did the chicken weigh?

4. On weighing a turkey, I found that I had a 5-pound weight, a 2-pound weight, a 1-pound weight, and an 8-*oz.* weight. How much did the turkey weigh?

5. A grocer, in weighing a roll of butter, put on a 2-*lb.* weight, a 1-*lb.* weight, an 8-*oz.* weight, and a 4-*oz.* weight. He said the butter weighed 1¾ *lb.*? Was he right?

Seventh Exercise.

WEIGHING WITH STEEL-YARDS.

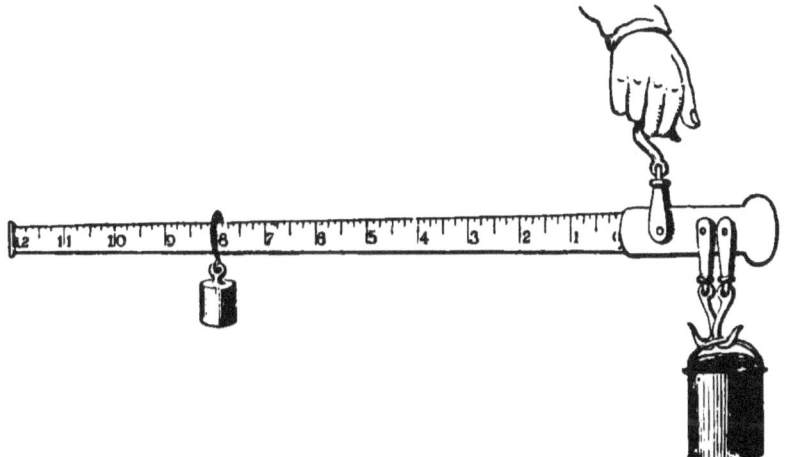

1. Here is a pair of steel-yards for weighing. A pail of butter is being weighed. You see that the pail is hung

on the hook nearest the large end of the bar, and the man holds it up by one of the other hooks. If you were to take the small weight off the long bar, would the pail of butter stay up? The small weight balances the large pail of butter just as a small boy can balance a large one on a see-saw. How is that? If you put the weight nearer the hook, how will the steel-yards act? If you put it further away than it now is? At what figure does the weight balance the butter? Then the pail and butter weigh how much?*

2. Count the large divisions of the bar. Each of these indicates a pound. Into how many small divisions is each of the larger divisions divided? One-eighth of a pound is how many ounces? Then each of the small divisions indicates how many ounces?

3. In weighing a package, I found that the steel-yards balanced when the small weight was at the middle mark between 5 and 6. How much did the package weigh?

4. How much does a package weigh which requires the weight to be 2 small divisions beyond 10 toward the end of the bar, to balance it? How much if the weight is between 7 and 8 and within 2 divisions of 8? How much if it is within 3 divisions of 8?

5. If your pail weighs 2 *lb.* and you want 5 *lb.* of butter, where will the small weight be on the arm when you have enough in the pail?

6. Where must the small weight be so that you shall have 4 *lb.* in the pail, if the pail weighs 1½ *lb.*?

---

* Doubtless a fuller explanation may be needed for many pupils, but the teacher can readily supply it, having the instrument before them.

Eighth Exercise.

WEIGHING WITH SCALES.

1. Here are three sorts of weighing machines. The upper one, on the counter, is the common grocer's scales. You can see them at the grocery-store. The things to be weighed are put into the scale-pan, and then the small weight is moved on the bar till it balances what is in the pan. There are figures on the bar just as on the steel-yards, by which you can tell how much is weighed.*

---

* If the teacher has no scales, it will create much interest and good feeling to take the little ones to the store where these can be seen, and let them learn how they work.

2. The scales in the left-hand side of the picture are called platform scales. The little girl who is standing on the platform of the scales is being weighed. The man is moving the small weight on the bar to find just where it balances, as you do on the bar of the steel-yards. How much do you think the little girl will weigh? May be your teacher can go with you to some place where they have such scales and teach you how to weigh each other. Such scales are used for weighing heavy articles, like barrels of flour, quarters of beef, dressed hogs, etc.

3. The other scales in the picture are called hay-scales. You see that they are just like the platform-scales, only larger. The platform is large enough so that a wagon loaded with hay can stand on it. The man stands at the bar to put the weight in the right place to make it balance. You see that the wagon is weighed with the hay. How shall the man find out how much the hay weighs without the wagon? If the wagon and hay together weigh 29 *cwt.*, and the wagon alone weighs 7 *cwt.*, is there a ton of hay?*

4. If the little girl on the platform scales has a package in her left hand which weighs 3 *lb.*, and the man finds that she, with the package, weighs 48 *lb.*, how much does the girl weigh?

5. If the groceryman puts up for me 5 *lb.* of sugar worth 9 cents per pound, how much must I pay him?

---

* Such questions which are a little in advance of the pupils' study should be thrown in occasionally to create or keep alive a desire to go forward and learn new things.

Ninth Exercise.

MEASURES OF TIME.

1. Here is a picture of a clock-
face. How many numbers are
there around it? Into how
many equal parts is the ring
around the edge divided by the
heavy marks? Into how many
equal parts are the spaces between the heavy marks divided
by the light marks?

2. How many pointers are there on the face of the
clock? Are both of the same length? Which is the
longer, the one which points to 3 or the one which points
to 12? These pointers are called hands. You can watch
the clock and see that the hands move around the face.
The long hand is called the *Minute Hand*, and the short
one the *Hour Hand.* To what number does the minute
hand point? To what number does the hour hand point?
Show in each of the clock-faces in the next exercise which
is the minute hand and which the hour hand.

3. Watch the clock in the room a little while and see
which hand goes the faster. Can you see either of them
go? Watch the minute hand awhile and see if it does
not go. Which goes the faster? See how many you can
count while the minute hand is going from one of the
fine marks to the next. Can you count a hundred while
the minute hand goes over one of these small spaces? It
takes it just 1 minute to go over one of these spaces.

4. Is a minute a long time, or a short time? Could you go home in a minute? Could you go to the door and back in a minute?

5. If it takes the minute hand 1 minute to go over one of the small spaces, how long will it take it to go from 12 to 1? From 1 to 2? From 2 to 3? How long to go from 12 to 6? How long to go from 12 to 3? How long to go clear around?

6. If you can have patience to watch, you will find that the hour hand goes from any figure to the next while the minute hand goes clear around. It is just an *Hour* while the hour hand is going from any figure to the next one. How many minutes in an hour?

7. How long does it take the minute hand to go from 12 to 3? What part of the whole way around is it from 12 to 3? How many minutes in a quarter of an hour?

8. How long does it take the minute hand to go from 12 to 1? How many times as long does it take it to go from 12 to 6? How many minutes does it take for the minute hand to go from 12 to 6? How many minutes in half an hour?

9. How many minutes does it take the minute hand to go from 12 to 2? From 12 to 4? From 12 to 5? From 11 to 12? From 10 to 12? From 9 to 12? From 8 to 12? From 7 to 12?

10. From midnight to noon the hour hand goes just once around from 12 to 12. How long does it take the hour hand to go from 12 to 1? From 1 to 2? How many hours to go clear around? Then from noon to midnight the hour hand goes around again. How many hours is it from noon to midnight?

11. From midnight to midnight again is called a day. How many hours in such a day? About how much of the day is it light? About how much is it dark?

12. If you have some kernels of corn in a cup and count them out one by one, picking them out with your fingers and laying them on the table as fast as you conveniently can, you will find that you can count out about 60 in a minute. If you count out just 60 in a minute, the time it takes you to count out 1 is a second. How many seconds in a minute?

13. Learn the following

TABLE OF MEASURES OF TIME.

60 *seconds* = 1 *minute.*
60 *minutes* = 1 *hour.*
24 *hours* = 1 *day.*

Tenth Exercise.

HOW TO TELL THE TIME OF DAY BY THE CLOCK.

1. At noon the hands are both together at the number 12. It is then 12 o'clock. One hour after noon the hour hand has gone on to 1, and the minute hand has gone clear around. It is then 1 o'clock. When the hour hand is at 2, where is the minute hand? What o'clock

is it then?   What o'clock is it when the hands are as on
page 145?   How long after 12 o'clock is 3 o'clock?   What

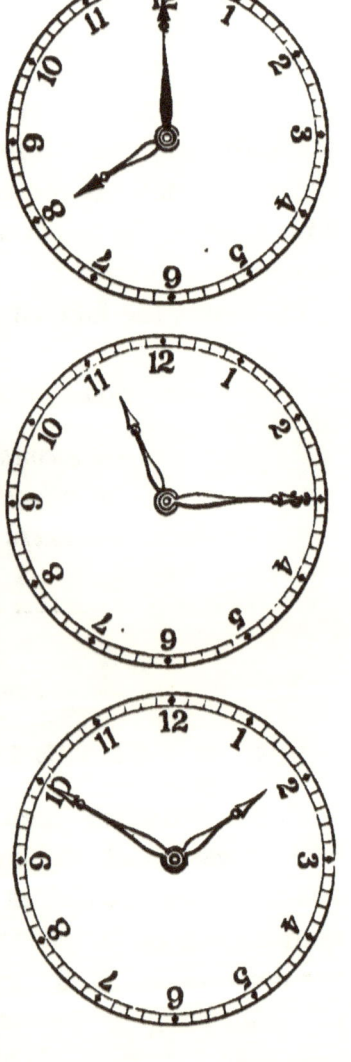

time is it when the hands are
as in this picture?   How long
is it from 8 o'clock on to 12
o'clock?*

2. What time is it when the
minute hand is at 12, and the
hour hand at 10 ?

3. Where is the hour hand
in this picture?   Is it before or
after 11?   When the hour hand
was at 11, where was the minute hand ?   How many minutes does it take the minute
hand to go from 12 to where it is
in this picture? How many minutes is it then past 11 o'clock ?

4. What hour is the hour
hand nearest in this picture?
Is it before or after 2 o'clock?
How long will it take the minute hand to get from where it
is to 12?   Where will the hour
hand be then? What time will
it be then ?   Then what time
is indicated in this picture?
How long before 2 o'clock ?

---

* These and the following are only given as specimen questions indicating
the manner of procedure.

5. What hour is the hour hand in this picture nearest? Is it before or after 5 o'clock? How long will it take the min-. ute hand to get from where it is to 8? How long to get from 8 to 12? Then how long will it take the minute hand to get from where it is in the picture to 12? Where will the hour hand be then? How long is it then before 5 o'clock?

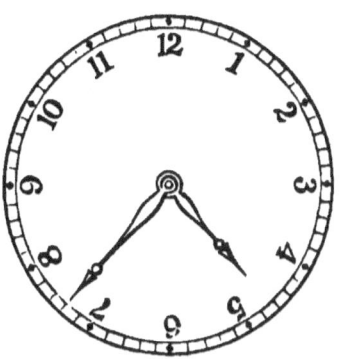

Concluding Lesson.

1. DAYS OF THE WEEK.—How many days in a week? What are their names? *Ans.* Sunday, Monday, Tuesday, Wednesday, Thursday, Friday, Saturday. What day comes after Tuesday? What day comes before Friday? What day comes after Thursday?

2. MONTHS.—A month is generally a little more than 4 weeks. How many months are there in a year? *Ans.* 12. Name the months thus: January, February, March, April, May, June, July, August, September, October, November, December. What month comes after March? What month after October? What before July? What before May?

3. The months are not all of the same length, but this little verse will enable you to remember the number of days in each:

Thirty days hath September,
April, June, and November.
By one more others vary,
Save the month February :
Twenty-eight this receiveth,
Until leap-year* one more giveth.

4. How many months have 30 days each ? How many have 31 days each ? Which is the shortest month ? How many days has the shortest month ?

5. Which month is usually just 4 weeks long ? How many days over 4 weeks do the long months have ? In how many places do two long months come together ?

6. January has how many days ?
   February has how many days ?
   March has how many days ?
   April has how many days ?
   May has how many days ?
   June has how many days ?
   July has how many days ?
   August has how many days ?
   September has how many days ?
   October has how many days ?
   November has how many days ?
   December has how many days ?

7. Now who of you all can tell how many $31 + 28 + 31 + 30 + 31 + 30 + 31 + 31 + 30 + 31 + 30 + 31$ make ? That is, how many days are there in a year ? This is a pretty difficult question for you, and you have not been taught in this book how to solve it. But the next book will teach you this and much more about numbers.

---

* The teacher should explain what is meant by leap-year—*i. e.*, in general, every 4th year.

www.ingramcontent.com/pod-product-compliance
Lightning Source LLC
Chambersburg PA
CBHW021110020726
47500CB00003B/689